This book grows hope and shines a light on the simple brilliance of life.
Chris Packham, author of *Fingers in the Sparkle Jar*

A book full of love, joy and a sense of deep reward.
Melissa Harrison, author of *All Among the Barley*

A moving unpretentious account of starting again.
Patrick Barkham, *Guardian*, Books of the Year

Beautifully written … intensely thoughtful and personal.
Helen Yemm, *The Telegraph*, Books of the Year

A very personal story of love, loss and rebirth.
Irish Times

A wonderful and moving book about how a slice of nature at the backdoor offers refuge not only to city wildlife but to the gardener too.
Alys Fowler, author of *Hidden Nature*

Quirky, passionate and endearing, an inspiring account of bringing a tiny garden back to life.
Dave Goulson, author of *A Sting in the Tale*

A beautiful story of a garden brought back from the dead.
Countryman

A glorious thing that is part autobiography, part gardening book and part fierce invective against the sterilisation of our urban landscapes.
Amateur Gardening

Made me itch to get out into my own garden and peer under piles of dead leaves to look for beetles. A moving tribute.
The Garden

It made me laugh. It made me cry. There is no louder, fresher voice for the value of urban wildlife.
Jules Howard, zoologist and author of *Sex on Earth*

A rallying cry for the wildlife garden.
Louise Gray, author of *The Ethical Carnivore*

Important and timely. I defy anyone who reads it not to want to do more to help their local wildlife.
Brigit Strawbridge, wildlife gardener and bee campaigner

THE BUMBLEBEE FLIES ANYWAY

A memoir of love, loss and muddy hands

Kate Bradbury

BLOOMSBURY PUBLISHING
LONDON • OXFORD • NEW YORK • NEW DELHI • SYDNEY

For Mum

BLOOMSBURY WILDLIFE
Bloomsbury Publishing Plc

50 Bedford Square, London, WC1B 3DP, UK

BLOOMSBURY, BLOOMSBURY WILDLIFE and the Diana logo
are trademarks of Bloomsbury Publishing Plc

First published in Great Britain 2018. Paperback edition 2019.

A catalogue record for this book is available from the British Library

Library of Congress Cataloguing-in-Publication data has been applied for.

ISBN: HB: 978-1-4729-4310-1
PB: 978-1-4729-4312-5
ePub: 978-1-4729-4311-8
ePDF: 978-1-4729-6126-6

4 6 8 10 9 7 5 3

Map on page 8 by John Plumer
Title page illustration by Jessie Ford

The quote on page 5, opposite, is from *Tom's Midnight Garden* by Philippa Pearce (2005), 21w from page 228, and is reproduced by kind permission of Oxford University Press.

Typeset in Bembo Std by Deanta Global Publishing Services, Chennai, India
Printed and bound in Great Britain by CPI Group (UK) Ltd, Croydon CR0 4YY

To find out more about our authors and books visit www.bloomsbury.com
and sign up for our newsletters

'And then I knew, Tom, that the garden was changing all the time, because nothing stands still, except in our memory'

Tom's Midnight Garden, *Philippa Pearce*

Contents

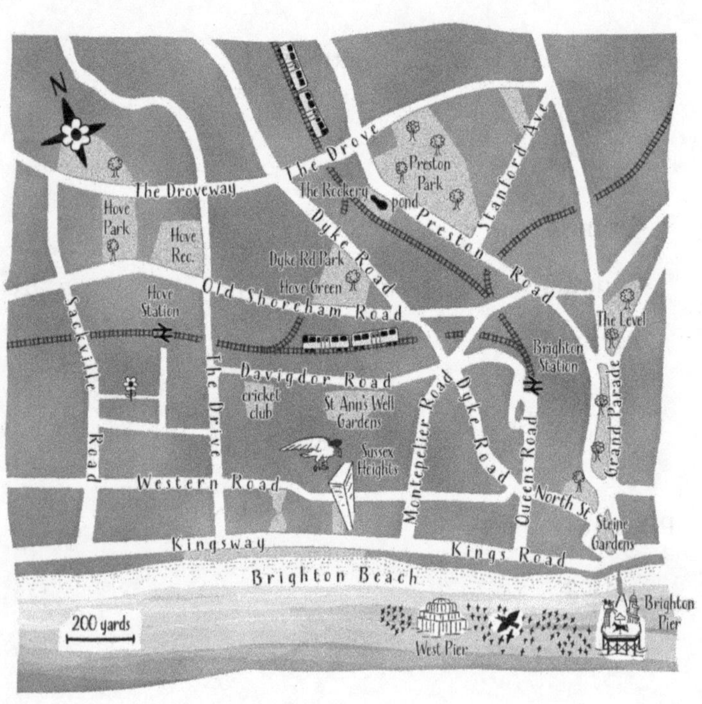
N
The Droveway
The Drove
Preston Park
The Rookery
pond
Hove Park
Hove Rec.
Dyke Road
Preston Road
Stanford Ave
Dyke Rd Park
Hove Green
Old Shoreham Road
Hove Station
Sackville Road
The Level
Brighton Station
Davigdor Road
The Drive
cricket club
St Ann's Well Gardens
Sussex Heights
Montepelier Road
Dyke Road
Queens Road
Grand Parade
Western Road
North St
Steine Gardens
Kingsway
Kings Road
Brighton Beach
200 yards
West Pier
Brighton Pier

PROLOGUE

A garden

In the suburbs of Hove, on a treeless street of terraced houses, lies a square of land where magic happens. It's where wool carder bees chase butterflies, house sparrows hang out with collared doves, red and blue damselflies catch flies while great fat bumblebees spill pollen and petals as they buzz from bloom to bloom. Where there are moths the size of your fist, where flies, aphids, caterpillars, slugs, snails, worms, centipedes and spiders reside. Where plants grow, flower and die – some of them to rise again and others to set seed before returning to the earth. Where compost is made, where birth, death and everything in between happens in a wild, unfathomable mess of struggle and pain and luck and fate. Where my heart beats.

It's not everyone's cup of tea. It's tiny, for a start. Six metres squared – barely a garden at all. North-facing. It's messy and wild, the grass tufty, the flowers self-sown. Some of them. There are piles of leaves and twigs and sticks and you can tell the trellis was put up with only one pair of hands. But it's all I have and it has all I can give.

Only I know where the sun shines. Only I know the spot where it hits the back of the fence in the last week of February, after months of darkness. The patches it lights up as it reflects across windows, the plants it streaks over in the course of the day, the year. Only I know the wet bits, the dry bits, the good soil, the bad soil. The plants, the plants, the plants. It's not a garden you read about in magazines, not something you'd come to visit. There's nowhere to sit, anyway. For you, that is. I hide in the deckchair squeezed between the pond and the climbing rose at the back. Gets the most sun. People aren't really welcome anyway.

It's a garden made from cuttings and stolen seed, dead turf and bits of root. Broken rules and a broken heart. It's all bird feeders and bee hotels. Things for the sparrows to eat, lure the goldfinches in. It's *noisy*! The pond takes up a third of it. It's ridiculous, really, but it's mine and it makes me happy. And, oh, before I came along it had been under decking for thirty years.

✿❀✿

When I don't have a garden, I go mad. In university halls I composted out of a window, grew herbs on communal stairwells, bought cacti at every student fair. In my second year I lived in a big shared house. Whenever the sun shone I would carry each of my sixteen house plants out of my housemate's bedroom window onto the flat roof and we would hang out together, like a sort of leafy teddy bear's picnic. I'd read or revise for an exam as we soaked up the sun together, while my housemate worked on the computer, just inches away from me, on the wrong side of the wall.

After uni I travelled for a year and came back with gardening books because I missed the soil. I'm a massive gardening bore.

It was inevitable, really, that I would end up writing about gardening for a living. That, despite degrees in politics and journalism, I would spend my evenings writing about plants, firing off articles to gardening magazines to see what happened. It wasn't hard to get a foot in the door – I'm twenty years younger than your typical garden writer. Twenty years less weary from repeating the same advice every year, twenty years cheaper. I joke that it's easy to be successful in a field no one else your age is interested in, that university graduates rarely take courses in horticulture and apply for jobs as garden writers, as I did fifteen years ago. But there's no other field for me. To make my job even more niche, even less desirable to my friends and peers, I write specifically about wildlife gardening. I harp on about the birds and the bees, the butterflies, the moths, the wasps. Please be nice to wasps, I write, in any magazine that will have me, every August. I glean statistics on wildlife declines, encourage gardeners to cut holes in their fences to enable hedgehogs to pass through, to plant early and late nectar for pollinators. I bore people about native plants, the importance of the food chain: the soil that feeds the plants that feed the caterpillars that feed the birds. I lecture friends on composting and mulch, alternatives to using peat, avoiding pesticides. I spend my days gardening or writing about gardening, and I will do this for the rest of my life. My friends think it's hilarious.

I tell people I go mad without a garden but they never believe me. It's an outdoors thing, I think. A sunshine thing. A plants thing. A looking-after-things thing. Still, with a garden I can be tricky in late winter and early spring – try to take me to the cinema or a bar when the sun's shining and I might cry and not be able to explain why. I might meet you for a drink and force you to sit outside in your hat and coat because there's a wisp of light behind the clouds. I don't mean to be difficult; it's probably a case of SAD. Mostly I arrange to see people after dark so they don't have to deal with me in daylight hours. See you after sunset, yeah?

Newly single after ten years, I was uprooted from my home and garden and nearly lost my mind. My books in storage, my plants in pots, my soul buried in some patch of earth I no longer had access to. I tried being a sort of nomad gardener; I gardened in an allotment, the shared house I lived in for a while, the gardens of friends who let me stay with them for a few weeks at a time. But it wasn't the same. You can't form an attachment to temporary things, fall in love, be so careless with your heart. Well, you can, but it only causes you more pain. And I try to avoid pain, on the whole, these days. Why plant things you will never see flower? Feed earth that will never feed you? I mowed the lawn in a house I was staying in and I ran over a frog. I had checked the grass but must have missed it somehow. Its guts littered the just-cut grass as I stood on the half-mown lawn bawling my eyes out. It was as if the garden was telling me to go inside, telling me this was the wrong thing to do. Better to avoid it altogether now and focus on other things. Like going to the cinema or something. Go to the cinema in the daytime, when the sun is shining and the birds are singing. Go and sit in the dark and suppress a scream.

I carried plants with me when I left: Japanese anemones, hellebores, snowdrops. They're not even my favourite plants. I think they were just a piece of home I needed to cling on to. I moved them all from place to place, along with a couple of grow-bags of tomatoes. Here she comes, they would say, with her little bag of plants.

✿❀✿

The day is Saturday. The month is January. The year is 2015. The sky is grey, with that thick blanket of cloud that hugs rooftops and makes you forget the beauty that can lie beyond it. You know that beauty: when the sky is endless, forget-me-not blue, cumulus clouds dotted here and there or stretched out like cotton wool, the blue backdrop filtered, changing. Then a bird flies overhead as the sun winks and everything is Promise. That beauty. Except today the cloud is oppressive: grey, full of rain but not ready to give it. There are no birds in the sky, no filtering blues and hues, no winking sun. The streets are lifeless and littered, bins overflowing, vomit and kebab wrappers telling tales on the previous night.

January. It should be a good month, a positive month, a month of out-with-the-old and in-with-the-new. Except it's not ever like that, is it? It's miserable. Everyone is miserable. I'm miserable. Hunched over on my bike, which I brought down from London on the train, I cycle these grey streets of grey vomit beneath grey cloud, looking at grey one-bedroomed basement flats where I can lick my wounds and start again.

Out with the old and in with the new, the unfamiliar, the frightening. I can't afford to stay in London. I weighed up the options in my head: Manchester? Too far away. Bristol? I don't know enough people. Brighton? Where I went to university and still have a few scattered friends. Where I can run along the beach, where I can buy a flat with a teeny back garden and still have room for a few pots out the front. More than I've had before if I can just tolerate two floors of people living above me. Can I? I'm not sure I have a choice.

I visit flats 'done up' for a quick sale. Flats with a hob but no oven, the bedroom looking out onto the street. Flats that smell of old dog and unwashed owner, flats with no bathroom at all, just a lean-to loo and shower tacked onto the kitchen. All of them are damp. There's one that almost steals my heart, the garden long and south-facing, but the ceilings are low and look ready to fall in on me. It's at the top of a hill so by the time I arrive on my bike I'm a half-dead thing heaving in lungfuls of other people's dust and mould spores. It doesn't feel right. Although nothing feels right. There's just one more before I give up. One more before I meet old friends in the pub and then haul my bike back on the train to London. One more basement flat with its dirty basement secrets. I cycle down the hill as the oppressive grey cloud yields its first spots of rain. January rain. Saturday-afternoon January rain.

It's 'open day' for potential buyers and I arrive to a swinging front door and young couples drifting, unimpressed, through the rooms. The estate agent waves me in, hands me the particulars. It's been reduced, she tells me, because of this patch of damp, here. The last sale fell through because of it, she says. The expectant couples turn their noses up. The flat is empty but light, airy almost, compared to the others, if you excuse the damp. I ask to be let outside and the agent fumbles for a key. I'm the first to have asked.

I step out into a gully of stones and dirty water, a mass of moss and weeds. My heart sinks. I climb the steps: decking, a broken fence panel, stones, cat shit. It's a desolate, barren wasteland – barely discernible as a garden. I pop my head over the walls on either side and there's more of the same: paved-over gardens, out-of-control buddleias. Ignored and overgrown or tamped down, suppressed. There's nothing for me here.

A quarrel of house sparrows chirps in a distant tree. Theirs is a fragmented habitat, of locked-away land with the occasional oasis of buddleia to sustain them. Where do they nest, find food for their young? I wonder how long they can hold on here.

Hope. Can you relate to a field of decking, a razed landscape? There's something about this desert and its doomed house sparrows that draws me to it. I want to unbutton the earth, let life back in. I wonder if I could save the sparrows or at least create a stepping stone from one distant tree to the next. When did a butterfly last land here? When did hedgehogs last roam? Can I restore this corner of Earth, this patch of emptiness? Perhaps. Perhaps we can fix each other.

Time was, this 'garden' was woodland. And woodland is trying to reign once more. A sycamore samara has long since landed in the stones and moss, where a sapling now grows. Ferns and buddleias have taken hold in cracks in the wall. The decking is rotting; plants are growing.

Nature has a habit of coming back to bite us on the bum. We can pour cement over a garden. We can chop down trees, uproot hedges, pour poison on 'weeds'. But they won't be defeated for long. At first everything will look as we left it. Razed. Cut. Broken. But then the healing begins: a button of lichen; a cushion of moss.

Woodlice will find the decking, and they will eat it. Spiders will find the woodlice, and they will eat them. Birds will find the spiders. Little by little, this barren, scorched earth will live again; it will defeat us. The decking will become rotten and slippery; moss will grow. The moss and rotten wood will provide a landing pad for the seed of a buddleia or ash or sycamore. Before we know it, we have a garden again. Blink and it's a woodland. And it's stronger and more determined than before, because only the strong and determined can survive here. Nature laughs at those who try to control it. It always, always finds a way.

I put an offer in. It's accepted.

Seven long months of sofa-surfing and waiting follow. I finally move in in summer. I decorate, unpack, sleep, cry. I wash and iron my clothes. I focus on the inside; the back door remains shut. I put up shelves, I restore a fireplace. I make bread and I service my bike. And then, one day, in early autumn, I venture out with my drill and start to unscrew the decking.

PART ONE

The bones, a skeleton

Autumn

What is this garden I have gifted myself, this unloved and broken thing? I drink tea at the top of the steps and lean against the wall, looking at it. It's like looking in the mirror: most of it lies a foot beneath decking, like the dead. It's sunny today and I'm pleased to see the space is well lit. North-facing, but there's nothing east or west of it – no tall buildings to block the sun. At different times of day I note where the sun meets the wall, which bit of wall, which bit of decking. I drink tea and watch it snake over the space; I have more to do than plot sun and land, but not much.

At the back, great willowherb, brambles and bindweed grow in the dust trapped among stones laid on membrane. Relics of farmland past or woodland perhaps, these colonisers are covering new ground without there even being any. And the sycamore – how many years must the stones have accumulated enough moss and dust for a tree sapling to send down roots? I tug it and it comes away in my hand. It hasn't penetrated the membrane, has rooted only in a few centimetres of dust and moss. There's hope for a world in which plants can grow in so little, and I feel bad for removing it. I check the willowherb for elephant hawkmoth caterpillars, the brambles for leafminers. Nothing. The space doesn't feel alive.

The decking is ancient, covered in moss and little clumps of purple toadflax which have seeded in the cracks; it's rotting in places. It also makes the garden higher, so when I stand on it I'm visible to neighbours three doors down. It sticks out like a sore thumb and sticks me out with it. I want to hunker down and hide, not be watched by a thousand neighbouring eyes. The design is mad: a square of foot-high decking laid on top of membrane. If you're going to put decking down, why make it so tall? Why not cover the whole space, rather than lay a margin of stones around the edge? This tiny space. Walled, and therefore closed to hedgehogs and amphibians. Without shrubs for birds to fly into, few flowers for bees to visit. My 'garden' is nothing more than a wooden box placed on top of something – earth, I hope. A box that, somehow, needs to be opened.

I lie in bed sifting through photos. There's a packet I keep in my bedside table and always have, wherever I've lived. It's an old 1980s Max Spielmann packet, containing a hodge-podge of photos old and less so – little windows into my first fifteen years. There are photos of me as a child: forced into a dress and placed in front of the hydrangea; a few years later in shorts in front of the holly; in the high chair; holding my new-born baby sister. There are rare photos of me with Mum and Dad, one of Great-aunt Gertie's ninetieth birthday, pictures of cousins, uncles, aunts, Grandad, random photos taken by me of a donkey and some cows. Mum with an acid perm (1980s); Mum in turquoise silk shirt, matching turquoise earrings and turquoise-rimmed glasses (1990s). And my most precious of all, the one I keep returning to and which is hard-wired into my brain, of the garden.

I was two when my parents bought a house around the corner from where my mum grew up. A four-bedroomed, semi-detached Edwardian 'cottage' in Solihull, it had servants' bells, beams in the ceiling and enormous wooden doors with a latch rather than a handle. It cost £2,000. It was a 'doer-upper', hadn't been touched for years. Spiders reigned and, whenever my parents' backs were turned, I would escape next door through the broken fence panel for a milkshake and a spider-free chat with the neighbours.

Backing onto the sports club that separated my house from the one Mum grew up in, the garden was a wilderness. I spent my earliest years looking out of the living-room window onto a 1950s patio with a hydrangea and a scrappy bit of cotoneaster, under-planted with lily of the valley. There was a long, thin lawn, a path that snaked between two ornamental borders and an old driveway leading to a concrete garage. A brick-built air-raid shelter, possibly one of the communal ones built by the government in the 1940s, served as a dark, ivy-covered shed. Beyond this was the vegetable patch, where foxes lived.

I don't really know how it started. There are photos of me aged two sitting in mud, planting and digging up things. I wasn't really encouraged, no one ever spotted the passion and tried to engage. I was left to my own devices, thrown out in all weathers while Dad checked the football fixtures and Mum cleaned the house. I made friends with the worm and the robin, the moth cocoon, the pigeon feather, and then at some point my rabbit, Mistletoe, who would roam alongside me, exploring. Two little fairies in the wilderness. Stigs of the Dump.

Now when I look at gooseberries I'm taken back to a time when I was the same height as the squat, spiky bushes, when I would watch, in detail, the veiny green fruit grow plump and ripen to wine red. Pick only the red ones, said Dad, the green ones are too sharp. The red ones were still always too sharp for me. Runner beans, raspberries, grapes. Fruit from a forgotten time, a 1980s childhood, the beginning of something.

The back of the photo says 'Summer '83', making it a year after we moved in. It was before my parents really worked on it but the vegetable patch is up and running. Dad usually took photos of the garden from the window of the room that was the dump and then my sister's bedroom and then his study and then simply 'the study'. But this one appears to be taken from a different room, the room that, at the time, was my bedroom.

I am eighteen months old and my sister, Ellie, is newly born; the garden is parched. Old Met Office weather reports tell me we had a good summer that year: June was changeable but mainly dry, July was mostly very hot and dry but thundery in places and August was warm, dry and sunny. A proper summer, not like we get now. There's little in flower but a few roses and the beginnings of crocosmia: early July, I should think. The photo is taken before the fence was fixed and I can see where my toddling self would sneak next door for a milkshake. The lawn is yellow, the borders overgrown and straggly. The air-raid shelter is not yet covered in ivy. There's a giant bamboo I don't remember but next to it is a tiny rhododendron that I would later claim as my own and climb. The 'rockery' is a mass of giant ferns; the garage has its door open and its windows reflect green and yellow back onto green and yellow. Still, now, I can close my eyes and stand in that garage and see chinks of light from those windows on the dusty floor.

Beyond it all is my dad's vegetable patch, where I played. There's the blue swing, netting covering cabbages and a huge row of runner beans. I count the canes: twenty-four, meaning forty-eight in total. That's forty-eight runner bean plants, as if anyone would ever need that many. Dad would send me through the leafy cane tunnel to pick the beans he couldn't reach. Go on, he would say. Just pick the young ones and pass them through to me. I'd enter this world hidden to all but the tiny, my chubby arms reaching above me to grab and pull fresh young beans from the plants. And the big ones, do you see any big ones now, he would ask when I thought I had finished. I always did. Pick them and pass them through as well, he would say. The big ones stop the little ones growing. I would pick those I could and alert him to those I couldn't, and his giant Popeye arms would reach in and up and pull them away. I was a good little child runner-bean picker, travelling through my row of canes with fresh green leaves and bright red flowers glistening in the leafy light. There were things here that no one knew about. Sticky gatherings of black fly farmed by ants and feasted on by ladybird and hoverfly larvae; slugs and snails waiting out the day to wreak havoc later in the coolness and darkness of the night. There were spiders here if I looked hard enough. It was a terrifying and thrilling experience in a world no one else could penetrate, where I was both lost and found.

Each time I look at this photo I see something different. Tonight I see the red T-shirt of the tennis player in the sports club beyond the garden, the roofs of houses built on the land beyond it, part of Mum's garden until Grandad sold half of it to developers. I see grey sky and a patch of white clover flowering in a parched lawn. I see the ghost of me in the runner-bean tunnel, the ghost of me in the garage, on the swing, on the lawn, in the rhododendron tree. My first ten years packaged neatly into a sun-damaged photo of a sun-damaged garden in 1980s Birmingham – the beginning of me.

I didn't have a BMX but I had a bike with BMX stickers on it. Like it was pretending to be a BMX, like I would think it was a BMX but it wasn't a BMX. It was a hand-me-down from someone or somewhere; I didn't have it new. I learned to ride on that bike, first with stabilisers and then without. Spokey Dokeys on the wheels; more stickers that came from cereal packets and magazines, stuck carelessly over the fake BMX ones. Siren whistle around my neck or in my mouth, I would cycle around the garden, from the kitchen up the drive to the garage, turn and then head back to the house along the winding gravel path between two large borders. Again and again and again. The path was too narrow for a six-year-old and a pseudo BMX bike, really. I'd take a wrong turn and fall off onto plants that spilled over the edge. Stop doing that! said Mum. I drove her round the bend. I'd be shoved out onto the street and told to be back for tea. Roam the streets that she roamed before they were streets. Before her dad sold part of her garden, before this Close was built here, these five houses there. When gardens backed onto fields and woodland, when you could sneak in and ride a pony called Tiny, fill jam jars with tadpoles and sticklebacks from the mere. Round and round on my bike, scratched legs and scabby knees, the ghost of Mum and her scratched legs and scabby knees around the corner. Houses, roads, sports club, a fenced-off mere. Carved-up land, locked-away play, out-of-bounds adventure.

When I wasn't on my bike, I was with Mistletoe. We would explore the garden together. He was given free rein – he'd sometimes sneak next door but he would always come back. Sometimes I would push him around in an abused doll's pram; later I was given a bright yellow harness, which I would squeeze him into and then 'walk' him to the shops, like a dog. I played with Ellie, of course. We would slide down the huge tree stump at the back of the garden, see how high we could get on the ancient blue swing. We made mud pies, expertly so, with different types of soil – clay here, stony stuff there. Add grass clippings and leaves, top with a pigeon or magpie feather, bird poo if we were lucky – the more gruesome the better. It wasn't a wild childhood as such. No one sat me down and introduced me to 'wildlife'. But we would glimpse foxes, watch blue tits in the nest box, gather blackberries from the mass of brambles at the back. We would find moth cocoons, all shiny and red. I didn't know what they were until I was in my late twenties. It never occurred to me to keep one and see what it turned into. We kept a stick insect in a jar once but I don't know where it came from. Bees and butterflies were sort of irrelevant to us, ignored. But I suppose it's easy to ignore things when they're abundant. Which, of course, they were so much more thirty years ago, although not compared to thirty years before that.

I climbed trees. There was a huge box tree beyond the kitchen, plus the two great rhododendrons, front and back. Ellie and I would dare each other to venture into the air-raid shelter with its ivy-darkness and mass of giant spiders among walls of old plant pots. I still shudder at the thought of it. It wasn't just Mum and Dad's stuff here but other people's stuff too. Dead people's stuff, war stuff: blackout curtains, the British flag. There was so much to explore and find, so much history I took for granted. I would escape into the garden and head as far away from Mum and Dad as possible. Make little dens in the long grass at the back, where no one could see or find me, sit in the greenhouse and smell tomato leaves, pick up moth cocoons and wonder what on earth they were, examine bright green gooseberries with their delicate veins and tiny hairs and the little residue of browned flower stuck to the base. I always had a stick to poke around in the soil with, hack at nettles and cow parsley with. A mud pie to ambush Ellie with.

Mum visits me in Hove, bringing with her bottles of wine, a few bits from home. We drive in her car to the Downs for a bit of country walking. It's a grey day, the first of the autumn mists lingers on the horizon; rain hangs in the air, waiting to pounce. Everything is quiet: there's no birdsong, not even a buzzard wheeling above us, and certainly no other people. We could have picked a better day, love, says Mum. We tramp on in our heavy, mud-caked boots anyway, watch squirrels cache acorns, blackbirds gobble rowan berries. We walk a loop that takes us on a quiet country road which, in summer, is flanked on either side by nettles as far as the eye can see, but which today is cut back, suppressed. Ashurst. It's the perfect English village: there's a medieval church, large houses and gardens, honesty boxes where you can buy chicken and duck eggs. We turn left into a farm where swallows and tree bumblebees nest in the outbuildings – memories of summer past. Everything we see today is eating or caching. Winter is coming.

Past the farm we take a path that runs between a copse of trees and barley fields, the woodland edge dripping with fruit. We walk among fallen leaves and fat worm casts, summer returning to earth.

I hate autumn. I can't get past the death, the decay. I see beautiful red and orange leaves and I want to cry. Everything is departing, dying or shutting down. And, yes, there are waxwings and other migrants that make winter a little less miserable, but on the whole it's deathly. Give me green and lush over grey and damp. Give me caterpillars writhing in the foliage, hedgerows bursting with birdsong, bees spilling pollen from flowers. As the half-life of winter approaches, masked for weeks by the Judas of autumn, I yearn only for spring.

The company is nice, though. Mum and I chat as we gather haws and sloes, a few sweet chestnuts, sparse compensation for the approaching misery. There's a chiffchaff in the great willow tree, having a last stab at holding its summer territory. Its call is weaker now, like it's finally giving up, too. We lunch in the pub before turning our backs on the grey autumn day, the decay and the giving-up, and head home.

In the garden it's been winter for thirty years. It doesn't change. There's no love, no life. The decking has paused growth of all but the strongest, and autumn is killing or putting to sleep those bold few anyway, just as in the wider world. Mum puts her feet up with a cup of tea and the paper. I'll just have a little rest, she says. I open the back door.

I climb the steps and sit on the decking. It's a mossy, slippery mess, patches of rotten wood where you'd go right through if you were careless enough. The walls are still clothed in the remnants of summer: brambles, bindweed, purple toadflax. The back wall is crumbling, neglect oozes from every pore. It's as if I've done this to punish myself.

Above me starlings heckle. They perch on the chimney tops making noises like R2D2, clicking and whistling and whooping. I feel like they're watching me, laughing at me. Sitting here, on this scorched earth, this suppressed land. It could feed them if I could just . . .

They would never dare come in here. Like the sparrows in next door's buddleia and the tits in the smoke bush over the wall, the starlings keep a respectful, safe distance. That locked-up land, that decking, is nothing to us, they sneer. I'm more determined than ever.

I don't know where to start. I don't know how to do this. Part of me wants to torch the decking but I'm worried I'll kill things living beneath it. And what lies beneath, anyway? Earth? Cement? Rats? I've no idea. Yet here, in this space, I might be whole again. I can feel it. I just need to find a way in.

I scrape fallen leaves from the surface, bits of dirt and debris, and I reveal screws. Screws! I can just unscrew it. Can I? I fetch my drill, some screwdrivers, a saw. You all right, love? says Mum, half-asleep on the sofa. I'm just trying something, I reply. And I set to, on this grim autumn day, spitting rain, death and decay. Laughing starlings. I start to unlock the earth, thaw the freeze, bring on spring.

There's something in the chimney. Something big. I lie in bed, alert, terrified of this thing, this movement, this irregular way to be woken on a Sunday morning. Is someone playing a trick on me? Is it a burglar? Will I see a foot? Are we in danger?

Quickly now it tumbles down, a great flapping crow. It lands in the hearth beside my bed, dazed, and then gets up and starts flapping and flying around my room. *Argh!* It's flying all over now, crashing into the window and the walls, pooing everywhere. I can't get out but I have to get out. *Muuum!* I can't walk or run or it will crash into me, chase me, poo on me. I wrap my duvet around my head and crawl out of bed and across the floor, prise open the door and run full pelt across the landing to Mum's room, panting. Mum, Mum! She's pulling her dressing-gown on. Darling, what is it? It's a crow, there's a crow in my room. She thinks I'm joking, thinks I'm the one playing a trick. Mum, there's a crow, you've got to help me.

Mum is now also terrified, and we have woken six-year-old Ellie, who stands pyjama'd in her doorway saying, A crow? A crow? Mum grabs a bottle of holy water and splashes it about as she says the Hail Mary. There's no time for the Hail Mary, Mum. We can hear it, crashing and skittering and shitting around my room.

Are you're sure it's a crow?

Yes.

A crow?

Yes!

What does it look like?

It's black and big.

Jesus Christ.

The three of us cross the landing to the scene of the intrusion, wiping holy water from our eyes. Wait here, Mum says, as she creeps into the room and we try to get a look. She closes the door behind her and from inside we hear laughing. It's a starling! She opens the window and lets it out and suddenly it's quiet again. Mum opens the door. She is laughing still but there's starling poo everywhere and I'm scared now that it will happen again and that anyone or anything could just let themselves into my bedroom via the chimney. I feel vulnerable, but also sad. I could have met the starling. I could have opened the window for it and maybe stopped it panicking and flying about. It could have been my friend. I pick a feather off the floor. It's long and brown, darker at the tip. I treasure it.

The starling is a wonderful bird. Sometimes brown, sometimes iridescent blue-green, sometimes spotted, depending on the time of year. Chattering always, able to mimic sounds such as police sirens and mobile phones. Sociable, doesn't take itself too seriously, likes dancing. I don't remember much of them in childhood, apart from that Sunday morning in 1989 when, aged eight, I came face-to-face with one I thought was a crow. But they are with me now, always, in the garden, laughing and chattering and whooping from the rooftops.

The starlings that whoop and whistle atop TV aerials and chimney stacks above my garden are the same that dance around Brighton Pier at dusk. They spend all day on the chimneys and then, a couple of hours before sunset, they become noisier. They seem to call to each other, egg each other on like friends texting before a night out. Impatiently, little groups fly from TV aerial to TV aerial, one or two slowly building up to six or seven. The whistles and clicks become louder, the flights become bolder, swifter somehow, as more join and the momentum builds, as they impatiently wait to get going. Then, suddenly, they're gone: whoosh! To the pier. Each little gang from Brighton and Hove's many rooftops and TV aerials, joining together for the big dance in the sky.

If you walk along the beach before sunset you can see them on their way. Their dark bodies bounce across the skies like charged telecom wires, throwing shadows across seafront buildings in the fading light. On the journey of two miles or so the starlings join up, meeting other gangs from other rooftops. Once, on my birthday, I was treated to a night in the Grand Hotel, made famous for the bombing during the Conservative Party Conference in 1984; on the fifth floor our balcony was at starling height, and we watched, champagne in hand, with the sun setting over the West Pier, as groups of ten or twenty or two hundred dashed to the party from all directions, calling and heckling to each other as they flew. Finally they came together in flocks of several thousand, moving through the sky in synchronised ribbons, contracting and expanding as one, a majestic being, a heartbeat pulsing above the city. If you stand on the pier and shut out the noise of arcade games you can hear the starlings whoosh past you, a million wings beating in symmetry, like wind rustling the trees. Stand close enough and on your face you feel the storm generated by their bodies. They roost in the bowels of the pier, beneath your feet. They never shut up, chatting and clicking as they rest, while their friends keep on going, refusing to let go of the night.

This thing they do, this dancing, is called murmuration. And Brighton Pier is one of the best places in the country to see it. No one really knows what it is or why they do it. Some scientists believe they do it to gain safety in numbers, to confuse predators before settling down to roost. Others think they gather to keep warm or exchange information. I don't care why they do it, I'm just glad they do.

Starling numbers, like almost everything else, are declining, probably, as ever, due to a lack of food and nesting habitat. In rural areas the loss of permanent pasture could be a factor, but in urban areas it could be the reduction of green space, the paving and fake-turfing of gardens. Starlings love to eat leatherjackets, the larvae of craneflies or daddy long-legs, which are considered a pest to many because they eat plant roots and can damage crops and make lawns look unsightly. Potent chemicals are used on the soil to remove them, which removes the food source for the starling and probably poisons everything else in the process.

Leatherjackets like lawns. In my last garden I would watch the females laying eggs, their pointed ovipositors easily negotiating the hard turf while blunt-ended males tumbled in the long grass. Perhaps, when this decking is pulled up, I'll plant a lawn and watch leatherjackets here. Anything to keep the starlings dancing in the sky.

Winter

No one ventures into their gardens in winter, hangs washing, chats over the fence. My only human contact is the twitch of a curtain or the opening of a door. A drift of cigarette smoke, a call to a cat. Otherwise it's just me, the birds and the cold, grey sky. All day, every day. After three months of unscrewing, chopping, bashing and cutting, the decking is finally up. The supporting beams remain, which are locked in place with cement, and around them is litter that was dumped beneath the decking when it was built, plus membrane and stones. The stones are big and heavy. They were placed over the membrane to suppress weeds, suppress life. They failed, of course. And now, the ultimate insult to those who lay membrane and stones: I have uncovered earth, black, crumbly earth, complete with worms and – somehow – plant roots. Willowherb and bindweed, I think, waiting for the black, heavy curtain to be lifted so they can grow again, live again, have their time again. Soon. Soon I'll be able to dig.

I make a sandwich and a cup of tea. I sit on a wooden post I've been unable to shift, which I later attack with an axe. My legs are battered and bruised from poorly aimed hammers and unseen screws, my fingernails split, my red nose runny. My sandwich is cold, my tea lukewarm. It's a celebration of sorts.

Early December, Saturday. The beginning of party season. I'm alone here, with mud and decking for company. Spots of rain assemble on my hands. Woodpigeons scoop their wings overhead, ploughing the sky. Wind tickles membrane and starlings laugh. Should I call someone?

My friends never see me gardening. In the old days, when we used to go out all the time and they would stay over, they might catch me in the morning sneaking outside while they nursed their hangovers on the sofa or blow-up mattress – a slug-shaped thing in a sleeping bag. What are you doing? Ssssh, stay there and sleep, I'm just out here. I would close the door on the stale hangover smell and the beer cans and kebab wrappers, and enter a world of bees and birdsong. A steaming cup of tea and pair of secateurs to ease in the day.

Other than that, nothing. We don't talk about it in the pub. Sometimes now, when they visit, they humour me: show us your garden then. And they say things like, Ooh, it's coming on, isn't it? Look at that wood. We return indoors and talk about music or film and I try to keep up. Occasionally someone will feign interest and I get to have half a conversation about it, like when I squeezed three people into my greenhouse and showed them my fasciated tomato flowers. How are your fasciated tomato flowers, they would ask, teasingly, in the weeks that followed.

We are still young, or so we tell ourselves, children of baby boomers. Few of us own our homes, least of all have gardens. Most of my friends are only just coming round to the idea that there's more than one type of bird. But they are coming round, some of them, slowly.

Where possible I hide it away, keep it to myself. I prefer it that way. Best done when I'm alone and the neighbours are out. Just me and the birds. Me and the hangover. Me and the soil and the spade and the overflowing compost bin that needs emptying and its contents sieving and spread on the borders. That's my favourite job. The dirtiest but most rewarding job. Shiny brandling worms glistening against black soil; compost so alive it's writhing. A beauty known only to gardeners, that I save for my alone time. I don't see friends when the sun shines.

The sky inks over and everything becomes still. A hundred herring gulls glide overhead, like fighter planes, towards the sea. Should I call someone? Starlings whistle to each other, forming small gangs to embark on their race to the pier. Even the birds are dancing.

On I press. There are so many bits of wood. So many little pieces full of nails, some of it rotten and compostable, others firmer, usable in some way. I chop up the best bits, bag it and leave it out for neighbours to feed their wood-burning stoves. With the rest I make my own flames.

Some of my best childhood memories are of roaming in my dad's part of the garden, his world of vegetables and cow parsley, of runner beans and raspberry canes, of the pumpkins he grew for Halloween, of cricket balls found in the undergrowth, which had come over from the sports club on the other side of the fence, of the swing I would stand on, one foot either side of Ellie, as we tried to reach the sky. Mum dealt with the ornamental side of things nearer the house: flowers, well-weeded paths, colour and orderliness. The back of the garden, the vegetable patch, was ours. It was a mess. But in the neglect there was magic. Here was a huge overgrown broom, a climbing frame. Foxes lived here, birds nested, I played. It was a wilderness, with long grass that towered above me, a clapped-out greenhouse sheltering a huge grapevine that had escaped through a broken pane like a triffid, and tomato plants that could take your breath away just with the smell of their leaves.

On summer evenings Dad would start a fire, and we, his two toddling daughters, would be allowed to stay up late to 'help' him. As he cleared ground for crops and piled waste onto the pyre, we would gather things to burn – a twig here and a fallen plant stem there – and we would throw them into the fire, without getting too close. The fire, a living, breathing, eating thing, with its roaring flames and trail of smoke that drifted off into the distance, was the highlight of our summer.

From our garden you could hear the Birmingham Superprix which, for just four years in the 1980s, raced around the streets of Birmingham on hot summer days. And yet, in this semi-urban habitat, with Red Arrows and airships flying above us, the din of the city just a few miles away, here was where I learned to love. Where I roamed, bare-chested and muddy in my outfit of green velvet shorts and oversized wellington boots, wielding sticks and mud pies, and where my dad, Ellie and I made flames. In 1980s Britain this was my access to nature. It was more than most of my friends had.

It didn't last. At some point my dad moved out and his overgrown vegetable patch was given over to wilderness. Between the ages of five and ten I tried, hopelessly, to restore it to its glory days of grapevines and gooseberries, hacking away at cow parsley with a stick, piling it up, asking Mum if we could have a fire (No), wondering why the weeds returned, thicker and faster, three weeks later, like creeping vines smothering our broken hearts.

It became so overgrown I was no longer allowed to venture there; my mum liked to see me from the kitchen window. It became a forbidden realm, like the air-raid shelter we could no longer get into, such was the impenetrable wall of ivy.

Eventually, my mum, sister and I moved to a small house a couple of miles away, with a tiny, well-tended garden, a pocket-sized path and a collection of shrub roses. And the people who moved into my former territory, who replaced me in My Natural World, razed my forbidden realm and put a tennis court over it. I know because I once sneaked into the sports club on my bike and scrambled up the fence and peered over. To see a world, my world, buttoned up, buried like that, destroyed me. That vision of my childhood lost beneath asphalt has never really left me. It may even have something to do with why I'm here now.

I start my fire. It's a damp, dark evening, with raindrops hugging the air. I build a small pyre, which I add to in the dying light as my fire replaces the sun and engulfs the garden. I add bits of wood, supporting beams the decking was laid on, and fetch twigs and fallen stems, which I throw on without getting too close.

✿❀✿

Scattered ruins reveal a life unravelled, a gardener without roots: kitchen scraps waiting for a compost bin, grow-bags of tomatoes I carried from place to place, bits and pieces I should have thrown out but kept for a rainy day. Three bee hotels. Stones, decking, mud; other people's litter.

The bee hotels. I haven't done anything with my bee hotels.

The concept of a hotel is a most un-bee-like thing. It suggests bees are as fickle and temporary as we are, that they want clean towels and soap wrapped in cellophane, dressing-gown and disposable slippers in the wardrobe. That they might stay for a few nights or a week, breeze in, breeze out. But they don't. They live there, some spend their whole lives there. There are around 250 species of bee in Britain. Just one of them is the honeybee, which makes honey and lives in enormous nests, mostly in hives managed by a beekeeper, though occasionally you find them in the wild. Then there are around twenty-five species of bumblebee, most of which make nests underground, in old mouse holes or under sheds; others nest above ground in long grass. All other bees are called solitary bees. Theirs isn't a matriarchal society like a honeybee or bumblebee colony; there are no enormous nests with a queen and workers and drones. Each bee is out for herself, solitary.

There are lots of different types of solitary bee but the ones that use bee hotels are cavity nesters. They've evolved to nest in holes in dead wood made by wood-boring beetles, and in the spaces in hollow stems that appear as plants rot away. But in our gardens dead wood is a health-and-safety issue, and broken plant stems are consigned to the compost heap. So a bee hotel it is – every garden should have one. South-east facing, morning sun. Can't go wrong.

Imagine a box filled with bamboo stems cut to fit the depth of the box and packed in tightly. A box of holes, essentially. This is your most basic bee hotel. You can use a variety of other materials – wood, paper straws, hollow plant stems, mud – but let's stick with bamboo for now. Having mated, a female will choose a bamboo stem and claim it as her own – from now on it's her nest chamber. She'll line the far end of the chamber with cosy material for her baby. Some types of bee use cut leaves or flower petals, others use mud. Then she'll gather pollen and nectar and make a sort of cake, which she will place within the lined chamber. Eventually she'll lay an egg on the cake and then seal the nest with more material (leaf, petal or mud). She has now created her first nest cell within the chamber. After this she will create a second cell, add another cake of pollen and nectar, lay another egg. By the time she's finished she will have filled her bamboo cane with a row of nest cells, each containing an egg and parcel of pollen and nectar. Each bee hotel can cater for many bees, all creating their own, individual nest chambers: red mason bees arrive in May and June and seal their cells with mud; leafcutter bees arrive from June to August and seal their cells with cut leaves. Blue mason bees come along sometime in the middle, and chew leaves into a sort of pulp. Use bamboo canes with different-sized holes (the smaller the better) and you attract a wider variety of bees.

I have three bee hotels: one consisting of an old champagne box filled with bamboo canes, another box filled with hollow stems of garden plants. The third is fancy – consisting of a purpose-built wooden block with nest chambers carved into it behind a Perspex screen, housed in a wooden box with viewing panels. You can open the viewing panels to see the nests the bees have made (best done in the evening so you don't disturb them). Trust me, it's better than telly.

Every summer I sit in front of my bee hotels and coo over bees. Red mason bees always; blue mason bees sometimes. Never leafcutter bees, sadly. At times, these bee hotels have been my world. I've lost days staring into them, watching the comings and goings, the carrying-in of a bit of mud. I know everything. I know which species uses which hole. Which are dummy cells designed to fool predators, and which are bursting with new life. These boxes of summer bring me such joy. I can't believe I've just dumped them outside with everything else.

My hotel of hollow stems is more successful than the rest. I made it last year; it attracted red mason bees within a week. I had temporary sleeping quarters on a friend's sofa; the bees had a permanent home in her garden. It's seen better days. Opportunistic spiders have spun webs in front of the holes to catch the bees unawares and it looks a mess, but it's been in continuous use for two summers. The bees must be attracted by the smell, it must stink of bee. My friend, Cara, lives only around the corner, so I brought the hotel with me when I moved here. I'm not messing too much with the ecosystem by moving bees a quarter of a mile from where they started nesting. And I can look after them better in my garden.

Except they've been outside since September, with the junk and the kitchen scraps and the wood and the stones. All activity has stopped now. The eggs that were laid in spring and summer will have hatched into grubs, which will have eaten the little cake of pollen and nectar left for them by their mother. In late summer they will have pupated, as when a caterpillar turns into a butterfly, their tough red cocoons protecting the grub as it metamorphosed into a bee. And these bees are just sitting it out in the box now, outside with all that rubbish, waiting, like me, for the sap to rise.

Most people leave their bee hotels outside all winter but it's a good idea to take them down and pop them in the shed to keep them cool and dry. If they get wet, they can develop fungal diseases, which can kill them. Also, tits and woodpeckers have learned that bee hotels are home to a great number of nutritious insects, a winter bounty, and they can raid them, destroying a whole generation, and weeks of hard work, in a matter of minutes.

I like to 'harvest' the bee cocoons and clean out the bee hotels. I do this in winter when the bees are fully formed inside their cocoons, so there's no danger of causing horrible mutations while they are metamorphosing. Harvesting helps to keep parasite numbers down, but I do it for the process. I get to count the bees, clean everything up and get it ready for spring. It's no more difficult than cleaning out a bird box. And, if I ever lived somewhere for more than five minutes, it would enable me to measure the growth in my garden's solitary bee population. You can plant flowers and create habitats, but only when you count each bee that's been laid in your bee hotels will you know if the population is growing, if the things you are doing are working.

I know, it's not for everyone.

It's dark already. I'm sitting on my dad's old sofa trying to watch the news. I'm feeling guilty about my bees.

There's no outside light. But it's time, I have to do this. I haul myself up, fire up my phone torch and open the back door. They're not hard to find. I grab the box of hollow stems first, followed by the other two. I brush them free of cobwebs and woodlice at the back door, bring them in.

I boil water for a night-time tea and clear a space on the floor – move the rug out of the way and pile two cushions in front of the settee. I make my tea and carry it over with the hollow-stem hotel. I tease out the stems and then gently pull the occupied ones apart, releasing the cocoons. There are woodlice everywhere, spiders; the floor is alive. But there are forty-seven bee cocoons, forty-seven parcels, each containing an adult bee sitting out winter, ready to emerge in spring. Most are red mason bees – the larger ones female and the smaller ones male. Some are blue mason bees. There are no leafcutters here, which makes me sad – I've never been able to attract them. I gently brush frass (bee poo) off the cocoons and place them in a bespoke 'release chamber', a little drawer purpose-built into the fancy hotel. This I will keep cool and dry in the porch until I have a shed, and then they'll live in there until the sap rises. Only in March will I fix it to the sunny back fence so the bees can wake up in the newer, smarter hotel, which is far less likely to attract spiders and woodlice, and hopefully return to nest here, rather than in the knackered old hollow stems they've been using for two years.

I replace the hollow stems in the box and put that in the porch as well. Just in case I've missed anything. I sweep up the woodlice, which have now worked their way to every corner of my living room, and take them outside again. My tea's gone cold. But the bees are OK. My bees are OK.

✿❀✿

A compost bin, two bare-root roses, an apple tree, a guelder rose, a spindle. Infrastructure. Native plants to bring in native insects – moths, leafminers and those that eat them. Creatures that will return, like the sun and the soil, to the garden that yet will be a garden.

I've been shopping.

I bought two terraced nest boxes for the house sparrows. They like to nest in loose colonies, so six homes in a row should be lovely for them. Still they gather in the buddleia and holly in the gardens on either side of me. They fly from one safe space to the other, avoiding me, the sea, in the middle. I'm the crevice between two mountains, the valley of death. Stay away from there, say the house sparrows. Stay away from that woman and her mud and stones.

Among the mud and stones is litter everywhere: old plant pots, bricks and builders' rubble left under the decking. Fag ends, chocolate-bar wrappers. It's endless. Every day I bag it up, take it down the steps, through the flat, out the front door, up the steps, to the bins. I stop, sometimes, when I find nice things to do something with, like the bit of wood I drilled holes in to make a bee block. Something might like to live in it. I take small breaks to dig and weed. Dig and weed! Each slice of my spade is a sigh of relief, a creaking of old bones, an intake of air. But there's *earth*, a whole chunk of it. Look at the soil, look at the roots. Look at it, look.

The soil is terrible. Dark and crumbly, a sort of silty clay, and it's loaded with builders' rubble. Tiny stones, bits of metal, lumps of earth, glass. I sieve it but soon realise I'm sieving out the goodness, too, those lumps of clay with all their nutrients. I make a note to find the nearest stables so I can bring life to this soil with manure.

There's no sun in the garden yet. It will be spring before it hits this patch of earth. I work in the shadows, knowing that when the sun finally comes, it will be the first the soil has seen in thirty years.

✿❀✿

I want to see my laughing starlings when they wake from their roosts beneath Brighton Pier. I've seen them dance at sunset so many times but never at dawn. I want to watch them celebrate the rising sun before they journey back to the roof of my house, heckling me as I unveil land that will feed them. The bastards.

Mid-December. The sun doesn't rise until after 7 a.m. but I want to be there sooner. My alarm goes off at six. It's cold. I dress in long johns and jeans, two pairs of socks, woolly hat and woolly gloves. I fill a Thermos flask with hot tea.

I step out of the flat and carry my bike up frosted steps. A blackbird gently practises his song from a rooftop aerial. Bin men work at the other end of the street, their creaking lorry lifting great metal bins into its mouth. I fiddle with my lights and then jump on my bike, away from the bin men and blackbird towards Sackville Road and the shore. I can barely see through the steam coming from my nose and mouth.

There are fewer cars than usual but still too many, the street-lit seafront already littered with dog-walkers and runners – I had convinced myself I would have this experience to myself. Still, there's a faintly nostalgic quality about it that reminds me of being on holiday, a newness I can't put my finger on. Beach flags ripple in the wind before the hidden sea. It's still dark but there's a faint green glow where the sun will rise, just east of the Palace Pier. My Star of Bethlehem.

There are three degrees of twilight. Astronomical twilight is barely noticeable, the first hint of day, when the sun is between 12 and 18 degrees below the horizon. Today it starts at 5.54 a.m.; I have missed it. Nautical twilight comes after, when the sun is between 6 and 12 degrees below the horizon, from 6.35 a.m. This is followed, at 7.17 a.m., by civil twilight: daylight to my untrained eye.

The wind hurts my face as I ride against it. The nautical sky is inky blue-black and I can just make out wisps of cloud and aeroplane contrail, as darker shadows of herring gulls rise up from the even darker water. The odd star holds on to the lightening sky but the moon is gone. Bright lights from the road advertise less salubrious activities: Casino, Envy, Legends.

I lock my bike to the railings by the pier, retrieve the flask of tea from my pannier and crunch pebbles down to the sea. I want to get under the pier, beneath where the starlings roost. I fail. I look up into the girders and see nothing. Sensibly, they roost where people can't reach them, further into the pier, above water. To reach them I would need a boat, or . . .

Seven bare-chested swimmers and a friendly Labrador step out from Brighton Swimming Club onto the pebbles. I sip hot tea, leaning against a rusty metal column as the swimmers scream into the water, dog splashing giddily around them. Some stay for only a few seconds while others swim out, beneath and around the pier. They could tell me where the starlings are.

The sky is lightening now as the sun inches towards the horizon, the gentle blue-pink of a perfect dawn, contrails and cloud reflecting the fire beneath them. Herring gulls line up on the railings and roof of the pier, clack-clacking expectantly: the starlings' alarm clock. Against them and the splashes of the encroaching tide I hear rising chatter. Two thousand birds waking up, getting ready to head back to the gardens and parks of Brighton and Hove.

Swimmers spill out from the sea as the pier ejects its first batch of starlings. The birds don't hang around but immediately disappear, a trail of smoke weaving into the distant sky, notes spilling off a sheet of music. I walk up off the beach and around to the other side of the pier and along Albion Groyne, a huge, brick-built groyne that takes me further out to sea and brings me closer to the roosts, the action. I wait, sip tea, laugh at herring gulls.

Civil twilight. They come out like bats now, thick black plumes snaking out into the day, contracting and expanding like a yo-yo as they head off in different directions. There's little dancing but who cares? It's just me here now. The swimmers are having hot showers indoors, commuters fill buses and cars beyond steamed-up windows. The pier is closed; this, here, is mine alone.

I watch my noisy, sharp-suited birds tumble out and disappear across the city as the golden glow of the sun rises from the horizon. By the time the sun is up they have gone and I cycle back along the seafront, following them, one eye on the sky.

At home I make tea and toast. I open the back door and greet the starlings on the rooftops, calling and chattering to each other as they have since they woke. I throw out mealworms and suet treats and retreat to the shelter of the doorway where I can watch them unnoticed, nibbling toast as quietly as I can. Still they laugh at me, still they don't come.

The thin, leaded window never fully shuts and a cold breeze blows through it. Dad has tried to block it with cardboard but it doesn't really work. We're watching the snow together, me standing on the windowsill, his hand on my back, steadying me. I am about three.

We're looking out to the garden, which has become a stranger in recent weeks. The snow falls in giant blobs and lands on top of itself as a blanket. It's proper 1980s snow, not like you get now; it sticks and builds and reinvents. There's no longer any lawn or winding gravel path, no border or driveway. Just snow from one fence to the other, only the trees and garage give the space away, identifying what lies beneath them. Dad points things out: the icicles hanging from the garage roof, the footsteps being gradually filled in. Snow mounts too on the window ledge and I wonder if it will reach the top and block the view. It won't, says Dad.

Dad lifts me off the windowsill and stands me on a chair in front of it. He starts taking photos and opens the window to get better shots. The snow comes in and hits my face. It sticks to me and melts on me. I touch it and it feels cold and gritty for a second and then it's gone. It's another world out there, he says, a world we don't venture into when it's like this, where things go on that we don't know about or understand. I ask him if we can make snowballs and he says yes, yes, just let me take these photos.

Movement, suddenly, at the back of the garden. It's a fox! Ssssh, can you see it, says Dad, can you see the fox? Bright orange against white, it's unmistakable as it slinks across snow. There's a den in the brambles at the back, we know this but we never see them. I've never seen them. Now it's as if Dad knew the fox would come out and we have been waiting here to see it. You must be very quiet or you'll scare it, says Dad. You must talk only in whispers and you mustn't bang on the window. Yes, Daddy, I say, as he puts a finger to his mouth. He changes lenses quickly, aims the camera, fiddles with the focus. He's poised to click and I bang. I bang really hard on the window, just as he told me not to. The fox bolts. Dad's furious, I can tell, but he tries to hide it. Why did you do that? he asks. I can't answer him. I suppose the excitement of being told not to do something was too much, or maybe I wanted to see the consequences of doing the very thing I was forbidden to do. Perhaps this is the perverse way children learn, or perhaps I am awful. I frightened the fox by banging on the window and my dad didn't get his photo. Still, I want to make snowballs.

Foxes come into my garden now although not because it's the perfect habitat. Clearly they can see beyond the mud and stones. I set camera traps to snap them – no one at the mercy of three-year-olds now. There must be a den nearby, in an abandoned garden perhaps or hidden beneath decking – I was surprised not to find one under mine. This one eats the peanuts I leave out for the starlings. I can't tell if it's a dog or a vixen but it's badly affected by mange. The camera traps aren't great in the darkness but I can still make out the rawness of its bare back and rump. Mange is awful but entirely curable. It's caused by a flesh-boring mite, which eats and defecates in the skin, causing it to itch. The fox scratches the itch and its hair falls out. Gradually it becomes so preoccupied with scratching that it hunts less, eats less. It becomes weaker and, in doing so, the mites take hold. The fox dies. Urban foxes are more likely to get mange because their diet is poorer. Those on a protein-rich, rural diet seem to cope better with mange than those that raid bins. I suppose it makes sense. This one is welcome to the peanuts if they can make a difference. Maybe one day I'll get a photo of it in the snow, for Dad, to make up for banging on the window.

In 1980s suburbia, foxes were a rare sight but they were all around us; we had to keep a brick on the bin lid to stop them toppling it over for the food scraps. Sometimes they'd break in anyway and we'd be woken in the night to the sound of a clattering metal bin lid; Mum would have to repack the rubbish in the morning. Foxes were a nuisance but nothing more. They lived at the end of the garden. Mum didn't like them going through the bins but generally we were happy to live alongside them. They were foxes and we were people and that was that. We lived in the house and they lived in the garden.

I feel sad when I see reports of them attacking babies, calls for them to be culled. So-called urban foxes are a different beast, or so the press will have us think: they're bigger, more aggressive, prone to breaking into homes, not bins, where they're more likely to feast on babies than discarded chicken bones. The truth is they're not bigger and they're not more aggressive – those entering our homes and attacking our children are more likely to be teenagers full of bravado, unaware of what they're doing. They live alongside us in the cities because we make it easy for them – you're never more than five feet away from a rat or a discarded box of fried chicken in the city. Easy pickings for a hungry fox, mange or not.

To see one out during the day was a rarity in the 1980s; these days, less so. They had a huge network of gardens to hide in when I was growing up; now I see them sleeping on shed roofs. Perhaps they're more visible now simply because there's less space for them to hide. Visible because their territories have changed and they're more likely to live in cities; visible because they're forced to live on top of one another, as so many of us are. I'd love to know where their den is. A typical urban fox territory stretches across eighty city gardens, apparently. How big is a typical urban garden? Mine must be smaller. I count the number of gardens back-to-back between the two main roads bookending mine: fifty-two. Fifty-two tiny gardens with a crossroads in between them. Is that enough for them? Where are the rest of them? Where's their family group?

I hear their screams at night. It's the beginning of mating season now, the blood-curdling siren of the vixen in heat, the owl-like hup-hup-hup of the male's response. I hear mating but I don't see it; I hear fighting but I don't see it. The camera traps pick up a bottom or a tail but these fleeting glimpses give little away. In some ways I wish I hadn't put the camera traps out, that my experience of them in the garden would be only through ears and an open window. The scratches on the wall as one jumps up to drop in; the shrieking, the fighting, the mating. Memories of a dustbin lid clattering on the driveway, a rare glimpse of one briefly before I scared it away.

✿❀✿

Winter solstice. Today the Earth is positioned in its orbit so that the sun stays below the North Pole horizon, as far south as it gets. All locations south of the equator have day lengths greater than twelve hours but here, far north of the equator, the day will last just seven hours, 56 minutes and 24 seconds.

I set my alarm for 6.30 a.m. and head out to watch the solstice sunrise. Christmas week. Schools are closed, many people have taken leave or are too hungover to care about making it into work on time. But it's still busy. There are always cars and buses and dog-walkers and runners. Even on a day like today.

It's miserable and I forget my hat and gloves. The wind batters my face, whistles in my ears. At the pier the sea rages and throws shapes high into the fairground rides. I pull my hood up and sip tea with my back to the wind. It's too much even for the swimmers, whose tactic in these conditions appears to be to lie on the beach and shuffle into the water, toe-first. They look like they're having a sort of unsatisfactory, angry bath.

I wait for ages. There's no horizon today, no nautical or astronomical twilight. No stars, orange sky or wispy contrails. It's grey and it will be grey for ever. There will be no sun. No rise and no set. I wait, drenched in sea spray, thinking of my warm bed, my home. Maybe I should return, it's the thought that counts.

As the sky lightens I notice little birds among the pebbles. At first I think it could be the starlings, that they, like the swimmers, have a different approach to leaving their roosts when the weather is at sixes and sevens. But, gradually, I see that nothing of them is like starlings. They are brown. They walk around in small groups with their heads down; they have a pretty little twitter, like a warbler almost, as they comb the pebbles. They are beachcombers. They walk close to the crashing waves and some are dusted with sea foam. Brown with white bellies. Turnstones, I discover later.

The herring gulls are not sitting clacking today. They don't need to wake the starlings, the wind and lashing rain will do that. Instead they shoot upwards on gusts of wind, like leaves being blown about, a bin being emptied of litter. Nothing settles.

It's miserable. I can't tell if the sun has risen. It's 7.20 a.m. and the starlings are late. They come, eventually, little ribbons snaking out from beneath the pier, hard to spot against the roaring sea and dismal sky. I watch them fly off, they as desperate to leave the seafront as I am. Still the sun hasn't risen.

Sunrise is supposed to be 8 a.m. I spot three people on the beach looking in its direction, other people braving the elements to welcome the solstice. It feels like a joke, they must be drunk, I must be mad. We all stand looking for something we won't see today. I wonder if I should join them and crack open a beer.

The sea has vomited out driftwood and other bits. Discarded fishing tackle, a tube of grease, an empty packet of crabsticks written in French. And, helpfully, boxes, crates, to put them in. If there's no sun there will at least be a beach clean. I start to fill the crates, one eye still on the pier, on my starlings.

The pier clock tells me it's 8.15 a.m. The solstice sun has apparently risen. It is still the bleakest of days, the greyest of December mornings. Yesterday it was beautiful. Tomorrow, maybe again. Is it the change? Must it not be smooth? Must we, who insist on looking out to the English Channel as one season gives way to the next, be so battered by wind and lashing rain?

Today all hope is lost. I fill the crate with fishing rope, a long-lost mop bucket, the French crabsticks, the grease. I find a dead, headless gannet, which I leave. Later, three crows battle the wind, and each other, for a piece of it.

After today the days will draw out, become longer. The sap will rise. I will stop this nonsense and return to the garden.

It's a robin that first comes to see me, as I dig a hole to plant the apple tree – the first thing I plant in the garden. She appears in a flash on top of the fence, flies to the wall and then spots me and darts off again. She's been here before, I realise, perhaps lured by the recently unveiled earth and its uncovered worms, or the compost heap, or even the hanging feeder with its mysteriously diminishing quantity of sunflower seeds. (Robins have only recently learned to use hanging feeders. And you can tell – they look ridiculous.)

I return to my work, digging away at my ocean of mud, lifting great fat worms. Look what you're missing, robin. I part-fill the hole with manure, compost and more worms, and fetch the tree from its bucket of water and mycorrhizal fungi. When I return she's back, standing stock-still on the wall with her head cocked. I stand stock-still clasping my apple tree, and cock mine. We eyeball each other for a moment, her presumably conducting a risk assessment on journeying into the garden with me in it, and me wondering how long I can stand still, holding a dripping apple tree, before forcing her to fly off again by moving.

Boom. She's on the spade. And then in the planting hole, picking at the worms and grubs before they retreat to safety. My robin. In my garden. My heart soars. I stand, rooted to the spot, watching her. I return the tree to its bucket and busy myself with other tasks for a few minutes, letting her work on the heap. Worms, in January. What a treat, little robin. And for me too – what better omen, that this, my first planting of my first tree, in this shambles of a garden, would be blessed by the arrival of a robin looking for worms. It's quite the ceremony.

Robins and gardeners go way back. As do robins and pigs. Take pigs into woodland and robins will turn up, waiting for the snouts to start rooting through the woodland floor and turn the soil to reveal worms, as they had for centuries before we started farming them. A gardener with a spade is nothing more than a pig, to a robin. The robin just needs to sit on the fence, on her little stick legs, and wait for her moment to sweep in and melt the gardener's heart.

Eventually I plant my tree, a local variety, 'Hawkridge', grown by Brighton Permaculture Trust. Bred on Hawkridge Farm, near Hailsham, in the nineteenth century, it was grown in orchards throughout Sussex when orchards were a thing. It bears medium-sized apples with a golden-yellow skin, pale red stripes and a crimson flush. The flavour is sweet and sometimes described as 'balsamic'. Whatever that means. Really I'm planting it for the red mason bees that roll around in its blooms in spring, the moth caterpillars that nibble its leaves in summer. The mistletoe that welds to its branches. And the robin that might sit in it and sing.

I firm the soil around the root ball, fashion a supporting post from a bit of wood from who knows where, which I push into the soil at an angle to avoid the roots, and loosely tie her in. She's a maiden, a stick in the mud, I can barely tell she's there. But she will grow and fill the space, eventually. It's a patient game, gardening. Next I plant two climbing roses – 'Shropshire Lass' and 'Frances E. Lester'. Both bear single flowers so the bees can access the pollen and nectar; no fancy double roses for me. 'Shropshire Lass' flowers only once, in summer, but produces good hips, bird food; 'Frances' flowers for months. 'Shropshire Lass' needs a lot of sun, 'Frances' can take a bit of shade. Both of them rampant, they will provide shelter for the house sparrows, if they come. And maybe, maybe, leafcutter bees will use the leaves.

'Shropshire Lass' goes in at the back, in front of the ugly south-facing fence, its planting hole part-filled with a little recycled manure, not really enough but all I have. 'Frances' goes to one side, at the top of the steps. They'll need support or trellis eventually but for now they're fine, these sticks of mine. I firm the soil around them, tuck them into bed. Hurry up and wake, I want to meet you.

Around them I plant teasels. Nine of them in various states of dishevelment. Like everything else they are all I have. I put them in, dot them around, imagine them all tall and cumbersome, the thick, spiky stems, the hedgehoggy flowers, the giant, veiny leaves. They look like little green blobs now, driftwood in a brown ocean, floating to a shore of a thousand pebbles.

✿❀✿

I call her Adrienne. I find her on the pavement in the dark, as traffic roars and raindrops fall. Amazing that I see her, black and velvety as she is. Adrienne.

I scoop her cold, wet body into my hands. Woken early or disturbed from hibernation, she will have been unable to find food. And here, on this bit of pavement outside Hove station, is where she would have met her end, either under someone's foot or just from exposure. But she's in my coat pocket now. She's coming home with me.

We walk back together, both of us shivering. I let us into my flat and it's colder still. I close the door, turn on lights, power up the central heating, find a takeaway carton. I pop her in, take my coat off, close the curtains.

Adrienne, it's January and you are not supposed to be in my kitchen. You are supposed to be in the ground, in the little den you dug yourself and lined with wax to make it waterproof. You are supposed to wake when the sap rises, when the crocuses flower, when the pussy willow hums. What an awakening is this, in a freezing basement flat? Huh? She sits, motionless, like a sulky teenager, a far cry from the queen she really is. Queen of the red-tails. I mix her a potion of sugar and lukewarm water, pop it in a bottle top and lower it into the box. Now, if she drinks it I've saved her. If not, there's no hope.

I can't bear to watch. I busy myself with other things. Give her a minute.

I've rescued grounded bumblebees before. It always happens in a mild winter. They wake too early but there's no food, and it's not long before they can't move. Bumblebees go into hibernation as early as July. They can go for months without nectar, tucked up in their little wax-lined dens. But when they wake, they need it. And they need it fast. They're always on the pavement or the path. I find them when I'm running, I find them outside train stations, on canal towpaths. Often it's raining, often it's the end of the day. It's as if they come out in the morning, spend the day looking for food and then realise their error so hobble to the nearest path so they might be saved. Bumblebee SOS. I like to think so.

Sometimes they warm up on the journey home and by the time you lay the sugar solution down they're halfway up your arm. You can see their antennae going, crazed as they are for their sugary hit. These are the best ones, the ones that want to be saved. They drink and drink, and within an hour they're flying around your home, trying to escape. But others, they can smell the sugar but it's as if they don't know what to do with it. They walk in circles, step in it, get covered in it. They limp around, sticky sugar on their wings. For years I put this down to parasites. That there was something messing with the wiring of the bee's brain. That it wouldn't, couldn't feed, a helpless, hopeless thing. What do I do with you now? Turf you out? I've made the situation worse. I've tried to wash the sugar off you, tried to warm you up. I've kept you for days encouraging you to feed. But it would result in only one thing: putting her outside on a mahonia flower, persuading myself that she wouldn't just die of exhaustion, that somehow she would be saved.

I've since learned that if you stroke her thorax she will drink. It can take a while, but gentle, soothing whispers and a little coaxing can work wonders. There, girl, suck it up. There you are, ssssh. They take longer to recover than the bee halfway up your arm, but they get there, eventually.

I check the box. She's drinking. Her proboscis is out, her tongue sweeping the bottle top from side to side, like a mop. Half an hour later and she's still drinking, still mopping from side to side. Oh Adrienne, my heart.

It's raining outside, still, but she doesn't want to stay. With a bellyful of sugar she's warming up, shivering her wing muscles, making low, deep buzzes. She'll have enough energy to find somewhere to rest for the night. But what about the morning? What if she can't find food then? I look forward to the days when there's a twelve-month supply of nectar in my garden. But now: what? I wrack my brains for the location of the nearest mahonia, winter clematis. Nothing. It's early evening, a good fourteen hours until the sun rises and she can start her search for flowers, or return to her hibernaculum. She'll have to stay, my reluctant buzzy guest.

I find the lid of the takeaway box and pierce holes in it. I remove the sugar water in case she knocks it over, add shredded paper. She raises her leg to tell me to back off. With the lid on she's cocooned in a dry space until morning. I pop her on the porch, where it's cool but dry. She'll be safe there.

Inside it's sad again. Unfamiliar, unhomely. I shuffle through my home, turning out lights. I cocoon myself in my little room. Like Miss Haversham. Like Adrienne.

In the morning I fetch her, growling and grumpy, a sulky teenager still. I pop her on the kitchen worktop for a cup of sugar – sit there, now, while I make this. I place the sugar water into the box and she takes a drink, not like last night but enough to see her on her way. I look her over as she feeds. She's in good nick, not like some queens you see after hibernation, all broken wings and worn coat, determined still to found her nest. However will you do it?

If bees could scowl. OK, Adrienne, OK. I stroke her thorax with my little finger, say goodbye. She raises a back leg. Pop her little box back on the porch, this time with the lid off and the sugar solution still there, just in case she wants to come back. And then I watch her go. She flies immediately, doesn't hesitate. Sometimes you see them orientate themselves, fly around in circles a few times to work out where they are and how they will return. But she doesn't do that. I saved her life but she will have seen me as her captor. No matter; she's on her own now and she has a chance.

Back inside and then out again into the garden. Funny how I didn't release her here, an unconscious decision but understandable I suppose – it's not good enough for bees yet, the first should come of their own accord. And will they ever? The state of it, really, a huge, muddy mess. It's not so bad at the back, now the soil is dug over, now there's an apple tree stick and climbing rose sticks. Now there are dishevelled, unwilling teasels. It will be spring again, it will be spring again.

Spring

It's a cool clear day. A Monday. Two leaves have burst from the stick stems of 'Frances E. Lester', two red-green flags waving in the breeze. The other rose, 'Shropshire Lass', is still in bud. The apple looks dead. I scrape a fingernail gently along a stem: green, I'm just being impatient. My nine teasels and a few hellebores are still driftwood floating in a mud sea.

Spring and there's so much to do, so much to be finished. And now it's cold, horribly cold, winter cold. But, March, the clock is ticking. I need to sow seeds, create and fill borders or I don't have a garden. But the space isn't ready, I've already failed.

The long winter is giving me time. But it's borrowed time, the other seasons will be shorter. It will be borrowed from spring or summer; I can only dream these weeks belong to autumn. Spring, when it eventually comes, will be a nightmare. Because it will overlap with summer and then –

Breathe.

Everywhere I go I'm on the scrounge. At Dad's allotment I dig up clover and comfrey. From Mum's I take bits of globe thistle and pulsatilla, two lady's mantle seedlings from a crack in the paving, a bit of bugle and granny's bonnet, a sliver of lemon thyme. I bag up a long stem of honeysuckle in water and treat it like a cut flower. With birthday vouchers I buy three-for-two oriental poppies, a mountain cornflower and a job lot of foxgloves. In a corner I find a forgotten pot of crammed-in sticks: common bistort and Japanese anemone, I think, which I gathered in a rush when I moved. And among the sticks a brown sodden leaf, a little scrap of lamb's ears. It's not much but I have things to plant among the teasels and apple, in front of the climbing roses. I have ingredients for a border.

I cut my honeysuckle stem into 10cm lengths and plunge them into a pot of soil. Everything else I arrange on the muddy surface and then plant in cold, useless earth, the globe thistle marked with a stick, the cornflower and poppies in front of it, foxgloves dotted variously. I try to ignore how pathetic it looks. I try to imagine them grown.

The sun at last has reached the back bit of soil and the plants sit in their mud baths, photosynthesising with little leaves. I lean against the wall and drink tea. While they grow I can set to work with the rest of the mess, the piles of stones, the litter, and by the time the sun fills that space maybe, just maybe, it will be ready for planting. But planting what? There's no plan. There's no design. There's no money. All I know is there will be plants in soil, plants grown from root and stem cuttings at the wrong time of year. Plants raised from seed during the coldest spring. Plants that might grow but could well die. Will there be a lawn? Maybe. I've some spare seed somewhere, no money for turf. Trellis, to add height, create privacy? Can't afford it. All I can do is sow seed and take cuttings, beg for little bits of this and that. See what happens. I have my teasels. There are always teasels.

But it doesn't matter. It's a garden. Plants grow and die, can be moved. Others will seed in, now there's earth for them to find. You can't have everything at once.

I return, half-heartedly, to the shade, to clearing stones, raking them here and there to reveal more soil. The rake scratches the stones and the ringing burns into my brain. I work like this for hours, until the sky starts to pink around the chimneys, the starlings gather and bounce their way to the pier, the herring gulls glide out to sea with the sun reflected on their bellies. In a distant tree the robin sings. I stop raking, bag up litter and rubble, take it down the steps, through the flat, up the steps and to the giant communal bin at the end of the road. I empty the bags into the bin so I can use them again. I make three journeys, carry six bags.

I take one final look at the garden. I need a shed. So much of my life is scattered here, pots, garden tools, bits of pond liner, slate and wood and an old piece of doormat I kept for a 'project'. If I could ever throw things away. With darkness waiting to pounce I drag the hose to the tap outside the back door. Connect it up. I turn on the tap and hear the water spit into the rubber, kicking it to life as I rush up the stairs to catch the end before it spurts everywhere. I stand, exhausted, watering plants into their new homes, giving them a life I wish I could give myself. I enjoy ten minutes of standing still, listening to the robin's dusk chorus and the clack-clacking of distant gulls. Darkness. The plants sated I turn in, turn off the hose, close the back door, run a bath.

But then it all comes crashing back: the loneliness, the What Am I Doing Here. This new home and city, this unfamiliar life, my things in wrong places. I feel sad, suddenly. Hopeless and useless. Without the garden there's no point to me. Without sunshine there's no life. The bath will be fifteen minutes. I'm a crumpled heap in the kitchen but I can't stop – not yet. I'm driving myself mad. My muscles are screaming but so is my heart and my heart is screaming louder. I retrieve the large oak dining table from behind my wardrobe, its legs from under my bed. It was Mum's, she wanted to throw it out, I wouldn't let her. It was the table that sat in the kitchen of the house I grew up in. That I had eaten my first dinners on, free from the constraints of the high chair. That I poured cereal over, and under, when I first tried to fill my own breakfast bowl. That I had been forced to remain seated at until I'd eaten every last morsel of food, night after night. The table which, as a teenager, I carved swear words into, while revising for my GSCEs. The table which I have no room for. The table which is now my temporary potting bench. My greenhouse staging. My me.

It's half the size of my living room. I move the sofa and other bits out of the way and assemble it beneath the south-facing window. It looks ridiculous but I don't care. I find a bag of compost, empty a kitchen drawer of seed packets and washed plastic containers, trays and yoghurt pots – any and all vessels. I assemble and turn on two heated propagators. My seed drawer is a mixed bag of packets stuck on the cover of magazines, seeds I'd saved from plants over the years, seeds bought from Gilbert White's garden five years ago but sadly, horribly, never sown. Most special of all, seeds given to me by Michael Blencowe of Butterfly Conservation. A packet each of caterpillar food plants for a couple of rare migrant butterflies: everlasting pea and fennel, to welcome me to my new home, with the scribbled note: *If you believe, they will come.* Butterflies. I screw up my face in excitement.

I turn off the bath and sow the seeds of my new garden. I sow my butterfly seeds first, and some other bits and pieces for various insects: sweet rocket for orange-tip butterflies, love-in-a-mist and cornflower from Gilbert White's garden, plus echinacea, *Eryngium* 'Miss Wilmott's Ghost' and a job lot of sunflowers and cosmos. Plus tomatoes, aubergines, basil and winter squash. I don't label anything – a final act of rebellion at the table I'd been forced to eat at as a child. I will regret this, of course.

Compost in seed trays is a happy sight and I feel better. I fire up the news on my laptop and watch it in the bath. I can't get the mud from beneath my fingernails, from the wrinkles in my skin. When I'm old seeds will sprout in my folds.

✿❀✿

I dream that my grandparents' garden is turned into a car park and wake up crying. I've not been there for twenty years: 1996, the year after my granny died, when we moved in with Grandad because he couldn't live by himself. I was fifteen.

It was a house without a number, simply called 'Driftwood'. Granny grew red geraniums in pots at the front, which she crammed into the porch for winter. They had a mad old Labrador called Sheba. Grandad manned the vegetable patch at the back, grew onions, marrows and stringy runner beans. Ellie and I would sneak behind the borders to hide; we played croquet, stole biscuits from the tin. The dog pissed great yellow medallions on the lawn and Grandad chased after her with a watering can.

Driftwood was a large, cold house with threadbare carpets and single-glazed windows. Heating was never allowed. I would come here on Inset days and make rock cakes with Granny or play In My Shopping Basket through the dumb-waiter hatch leading from the kitchen to the living room. She taught me about the dawn chorus here, one magical morning where she terrified me with birdsong. We walked Sheba down country lanes where, once, we heard a cuckoo. We ate panda-bear-shaped meringues from the local bakery, watched *The Six Million Dollar Man* and *Fifteen to One* on the telly. Granny would feed the birds and we would watch from the kitchen as house sparrows and tits battled it out for stale bread and bacon fat. Sometimes she would treat me to a tumbler of boiled sweets.

I wish I had known her better. About ten years ago I started grieving for her – she died in 1995 but it wasn't until much later that I appreciated the loss. Appreciated the relationship I could have had with this woman who had lived through the Second World War and lost a brother, who married late and raised four children, nearly lost everything when Grandad, a wine merchant, lost his shop during the recession in the 1970s. Who muddled through and built a life back again, cooked dinner and ironed shirts for her husband, who drank gin every day and had regular blackouts. Who I knew only as the woman who knitted me jumpers and made me rock cakes, taught me about birds and came to school plays when my parents were too busy. How wonderful it would be if I'd been able to talk to her about her life, if she had been there to guide me into adulthood, if we had got to know each other beyond rock cakes and jumpers and the cuckoo and dawn chorus.

She died around the same time I realised I was gay, and in her last moments I distanced myself from her because I feared she would reject me. A little old lady in a fireside bed, me awkward and ashamed beside her. The last thing she said to me was be quiet, close the door, stop making so much noise. We moved to Driftwood after she died. But it wasn't the same. She wasn't there, for a start, and neither, really, was Grandad. I would come home from school to find washing-machine parts in the bread bin, food for the evening meal mysteriously disappeared. He would answer the phone to my friends and tell them he'd never heard of me. The setting was more rural than I was used to, the last train to Birmingham was at 6 p.m. – you had to put your arm out to make the train stop – and I would head out to the bright lights of the big queer city, blissfully underage, knowing I would not be able to get home until Monday. I don't know how long we lived there for – six months, a year maybe. I slept in the bed Granny died in.

I never gardened at Driftwood but I harvested runner beans, played on the lawn with my cousins, terrorised the dog, watched birds. There were so many birds. As I grew I would be tasked with jobs: raking up leaves, clearing windfall crab apples so Granny could make jelly. Jobs that involved getting cold wet sleeves that would creep up to your elbow and never dry.

In my dream, the driveway snakes around the house, taking up both the front and back gardens. I cry for the birds Granny made me listen to and which scared me so much in 1989. I wonder if they are still as noisy now. I drink tea in bed and open the laptop, find the house online, last listed for sale five years ago but unsold. The front garden and drive are much the same, the crab-apple tree still standing, the huge bank of conifers separating the house from next door. The little piece of wood, on which 'Driftwood' is carved, still pokes out unevenly from the lawn.

The back garden is not a driveway either but the patio has been extended and the right-hand border is gone. Most of the shrubbery remains but there's no vegetable patch at the back, no tired shed, no compost heap. There's no bird feeder; there are no piss medallions. I see the ghost of Grandad asleep in his shorts in a deckchair, skin raging pink from the sun. Croquet balls litter the lawn, Granny's washing on the line.

✿❀✿

Leggings and shoes, base layer and jacket, ear muffs and woolly hat. Temperatures are well above freezing now and there's been rain. Brighton and Hove's amphibians are on the move.

I wait until after sunset, after the starlings and herring gulls have left the rooftops for the pier. After the sky has flashed and died, after the stars and the moon have risen. As I head into darkness, spots of rain hit my face. I bunch my hands into my sleeves, feel for the zip of my jacket in case I can tweak it up a bit, kid myself I can be warmer.

My headphones are in, dance music to pump the blood, to ease me into the slog of running in cold rain, the hills, the night. Haloed streetlights guide the way as I reluctantly start, weaving past commuters heading home from London, up past the chippy and Hove station, the sorting office and Tesco Express. Down again and left onto Cromwell Road, dogs on leads, buses splashing puddles as the heavens open to refill them. Good for frogs, I suppose. I run faster to keep warm, left again at the lights and up into the rain-soaked hills to join the Droveway, an ancient route that used to link the parishes of Lewes and Portslade but now connects the parks of Hove and Preston. In times gone by farmers would drove cattle along here, to or from market or between summer and winter pastures – maybe frogs and toads would have used it, too, to find their mating grounds in spring; there must have been dew ponds here where houses, another chippy, another Tesco Express now stand. Garden ponds now home to the descendants of simpler times, perhaps. If there are any. A dairy still stands, its buildings once part of Preston Farm. These, and a few old houses on South Road, are the only hint of the Droveway's past.

I come out at the bottom of South Road by the petrol station, turn right along Preston Road until I reach the Rockery. It's always open, closes only for Pride. Right on the main road but a world away from it, it used to be part of a wood called the Rookery, where rooks nested. It's a magical little place, three and a half acres of pathways, pond, stream and waterfall, all surrounded by wonderful plants. It's managed by one staff member and volunteers by day, used by teenagers at night – sometimes, on morning runs, I clear away empty beer cans and crisp packets so the park volunteers don't have to. There are goldcrests and firecrests here, song thrushes that drown out the sound of traffic on spring evenings, a wren that nests in the gunnera. On hot summer days terrapins bathe.

It's too cold for terrapins and teenagers tonight. I stop running, sweat and rain clinging to my skin, breath steaming around my head. At least the rain has eased. I'm alone but buses and cars are just metres away – no one cares to see me. I fumble for my phone and turn on its torch, walk to the edge. I'm a lighthouse, my searchlight scanning the sea. Ripples in the pondwater could be frogs or fish, waving not drowning. Great orange carp surface to greet me.

Soaked as I am I crouch down into wet grass. In the din of the traffic I can just make out the low rumble of croaking frogs, like the faraway sound of a woodpecker pecking. Little growls and squeezes, ssssh. I walk around a bit to a shallow shelf almost separate from the rest of the pond and here I find them gathering for their party. There are hundreds of them, my torch lighting up an ocean of eyes. Great fat females and blue-stained males, couples in amplexus, the odd three-way. But it seems remarkably sedate, really. I've imagined frog orgies as a big mating ball, a writhing mass of rubbery bodies. But here, now, it seems like a waiting game. A bar in early evening perhaps, before the drink flows and inhibitions loosen. The males appear to have stations, each one a little way away from the next, legs splayed, croaking and waiting. What next? Does the party start later? Is this the party? I watch them for a while, sometimes torch on and sometimes torch off, I don't want to disturb them. The low growling croaks become longer, the splashing more frequent. I don't know what I'm expecting to see but there's no spawn yet and no obvious signs of it being produced. The frogs are sitting it out but I can't any longer. It's raining again and I can't feel my legs.

I leave the frogs, running as fast as I can, back up the hill and down Cromwell Road. Back past the Tesco Express and Hove station, the chippy and now late London commuters – poor buggers. Inside is much the same. Warm but . . . I run a bath and think of frogs, legs splayed, waiting.

I ask Mum about her childhood garden and she draws me a map. It doesn't have any detail to it but she adds surrounding gardens, the road, puts her childhood in the context of mine. She draws the field at the end that she and her siblings would break into to ride the pony they named Tiny, the house where the priests and bishops lived, including the shed housing a homeless man called Harry. Granny would make him lunch every Christmas, says Mum, and send her little brother Pete to deliver it on a tray with a glass of beer. She draws three neighbouring gardens, the sports club, the mere. Not my garden. My garden didn't exist in her childhood but I place it instantly. There was just one neighbouring garden between hers and the sports club. If Grandad hadn't sold part of theirs to developers I would have been able to glimpse her childhood from mine. Seen its trees, by then twenty years taller than when Mum knew them, climbed them. I think of the photos Dad took from out of my bedroom window, the image of the house roofs beyond the cricket pitch, the people who lived where Mum used to play. People need houses. But can one roam from Mum's old garden all the way to the woodland around the mere, as children did then? And not just children but wildlife, too. Hedgehogs, toads, frogs – how far can they roam in new estates, land carved up and fenced off? Children had wilder childhoods and wildlife had it easier. Will we ever get that back?

She was nine when they moved there, in 1963. There are albums of grainy colour photographs of her and her siblings at various ages: sitting together on a picnic blanket, sunbathing, bikes and outdoor play. She spent her teenage years here, breaking out of upstairs windows to go to gigs, throwing parties when her parents were away, drinking wine stolen from Grandad's cellar. It was sometime in the early 1970s that Grandad and two of their neighbours sold the back portions of the gardens. Mum doesn't remember the date or if Granny approved; she may not even have been consulted. They sold the house in 1979. I find it online, last listed for sale a couple of years ago but frustratingly with no photos. Next door was listed recently but has only one photo; the house three doors down has twelve photos and an enormous garden – not one of those that was sold off. Was Mum's garden like this? It's huge, magical. It has a patio with statues on either side of steps leading down to a giant lawn. The statues look like lions. Did Mum have lions? Mum's house and garden remain a mystery, known only to me via stories from her and her siblings now in their fifties and sixties. There were, sadly, no statues of lions.

✿❀✿

My plus-size potting bench stays in the living room for weeks, supporting leggy plants and puddles of muddy water. Compost and fungus gnats reign. I'm the crazy plant lady with a flat full of flies and a garden full of mud. We don't answer the door. We talk only to each other.

Outside I dig and shift stones, dig and shift stones. I go running still, up into the hills and along the Droveway to the Rockery, where I stop, every day, to look at frogspawn that's now been laid in enormous clumps. There are clumps on clumps, a mass of gelatinous goo. It's beautiful.

And then I stop talking to the flies and I stop shifting stones and I start digging a pond. I didn't plan to. My garden is too small for a pond, too shady. In digging one I'm creating more work for myself, more stones, more rubble, more mess, less time. But I don't care. I don't even know I'm doing it. I couldn't tell you when I realise, but I do, firmly and passionately, that I never want a garden without a pond again. I love them. I love the hum of insects swarming on the surface on a summer's evening, the fleeting glimpse of a waterboatman surfacing for air, the ridiculous tail-chasing of whirligig beetles. I love the plop of a frog disappearing into the water as it hears you coming. Tadpoles. Huge, alien-like dragonfly larvae. I could sit and stare into a pond for hours. And I do, I stop and stare into one every day when I go running. The fact that, even at just a metre in circumference, a pond will take up a third of my stupid muddy garden doesn't matter at all.

Besides, I have a spare bit of pond liner. And some pond plants. And a million stones.

We didn't have a pond in the garden I grew up in. I ruined the chances of it ever happening after I stepped on a lily pad in a garden centre aged six, while my mum filled her trolley with bedding. I'd just learned that Jesus had been able to walk on water and I wanted to see if I could too, given that I said my prayers each night and went to church on a Sunday. So I stood at the edge of the pond and carefully stepped on the lily pad, gradually increasing pressure until I inevitably fell in.

The water wasn't deep. Waist height, I was lucky. But it was smelly and there was lots of pond weed. I felt silly. I stood, semi-submerged, for what seemed like ages. Eventually, Mum, alerted by four-year-old Ellie, pulled me out and I was wrapped in a bin bag so as not to spoil the upholstery in Dad's car. Someone gave me a packet of crisps. Mum left her trolley of plants behind. Years of guilt-induced swimming lessons ensued, plus a refusal ever, ever to have a pond in the garden. I have always blamed Catholicism.

My life is a patchwork of temporary things. Temporary love, annuals. I've grown plants I've never seen beyond their first year. Dug how many ponds? Three. Some roots would be nice. Could this pond, this fourth I'm digging, be the one I see frogs spawn in?

It's going under the wall, in front of me as I climb the steps from the gully into the garden proper. It will force me to walk around it to get to the back of the garden so I don't go in a straight line, make the garden look bigger. It should get a bit of light at midday and then more in the afternoon as the sun swings past the chimneys above. It will have a circumference of 1.5 metres, a maximum depth of no more than 30cm, and a gradual, sloping edge where invertebrates can congregate and where amphibians and mammals can enter and exit easily. Birds might bathe in it, dragonflies might breed in it, frogs might spawn in it. I have high expectations for my tiny water.

In the books, there are various instructions on how to dig a pond, starting with marking out the shape before you start. None of them says start digging and see what happens. But that's what I do. The revolution is sort of potato-shaped.

It doesn't take long. I don't want to dig too deep, most life is in the shallows. I smooth down soil, create shelves for plants, level it off so the top ridge sits below the garden's soil level so I can bury pond liner beneath stones. I remove as many pebbles and bits of glass from the soil as I can, lay down soft liner and then butyl. I manage to cover two stubborn lumps of cement the decking posts had been sunk into. Nice, that. That the tools designed to suppress life are, in turn, suppressed by those that will give it back.

I have, of course, a bucket full of water and pond plants – water forget-me-not, frogbit, pennyroyal, pickerel weed. Left over from a photo shoot a year ago, brought back on the train, kept alive somehow. Here she comes, with her bucket of plants. I divide them into separate planting baskets, trim them down, arrange them on the shelf. The water they were sitting in goes in the pond, along with some pond snails that seem to have hitched a ride. The rest of the water will come from the sky. And the stones, the endless mountains of stones, suddenly have a purpose. I sort through them, finding the prettiest to place over the pond liner. My pond. My beautiful pond.

I drink real ale with my new friend Rachel in a micropub that's opened up by the station. We met through a mutual friend in summer, where we chatted, suddenly and unexpectedly, about our shared, and slightly macabre, interest in death. She attends death workshops and I collect the skulls of dead animals. We also enjoy real ale and gardening – I've built friendships on less in the past. We sit on high stools and work our way through the menu of porters, wheat and rye beers, ales made with local hops. She entertains me with stories about soul midwives – people tasked with guiding you on your journey from life into the afterlife; I tell her about the fox, badger and hare skulls on display in my living room, the mole and goldfinch in my freezer. I don't know what to do with these, I tell her, they're so beautiful. I think I might learn taxidermy. She's the only friend, so far, to empathise with my predicament, to not tell me to stop talking. We drink absurdly strong porter, play cribbage and chat about our gardens; hers backs onto the allotments where, a couple of years ago, the rare long-tailed blue butterfly, which comes over occasionally from France, was found breeding on everlasting pea. I will give her some of the everlasting pea seeds Michael Blencowe sent me so she can grow them and keep her eyes peeled. She'd like that, she says.

At home I'm slightly giddy and I don't want to go to bed. The new shed, which I bought online a while ago, has been leaning against the wall in the hallway for a week, the smell of fresh wood a constant reminder of my tardiness. Tall and thin, with two 'eyeholes' at the top of the front panel that I imagine robins might use and make a nest inside, it's called a Sentry Shed. It's a small thing, tall and slim. I can put it together while catching up on the news. It's only 11 p.m. I fetch my drill and some screws, haul pre-made wooden walls into the living room, turn on the TV, ignore the How-to-build-your-Sentry-Shed instructions.

It's not hard to do but I probably shouldn't have embarked on the task at 11 p.m. on a Friday night, after one too many 7-per-cent ales. And it would have been much easier, and the result less wonky, if I'd had help. The 'flat-pack' consists of a back and front wall, two side walls, a floor and door, hinges and latch, two shelves and two roof panels. Putting them together is just a case of working out which bits to fix together in the right order, and ensuring the roof won't let water in when it rains. The assembly takes a few attempts, especially the alignment of the roof panels. But I manage it, eventually. The job done, I stand back, admiring my handiwork, my pretty Sentry Shed.

I go to bed sated, having completed one more job that will make a huge difference to the garden. I fall asleep thinking of the things I will store in it, a shelf for paint, perhaps, and another for scraps of wood I've saved to make bee hotels. I'll fix screws to the inside of the door to hang tools, put a padlock on the latch.

In the morning I wake to find my pretty Sentry Shed is too heavy to move alone and taller than the height of the back door – I'll have to tip it to carry it out. It stands, teasing me from the centre of my living room, where it will now have to stay until someone comes round who can help me move it. I wonder if I can persuade the postman, Keith, to help, or if that would be inappropriate. I don't know when anyone is next coming over, the place isn't yet ready for visitors. I am an idiot who shouldn't be allowed to assemble sheds after drinking ale and chatting about skulls and soul midwives. I am an idiot with a shed in my living room.

✿❀✿

I'm all alone in my half-made mess, my plot of mud where nothing grows. The mud and the cold, the slow grind of the rake. When will it end?

Like a robot I rake stones, pick up stones, put stones in a bucket, which I carry through the flat to empty over the ugly cement at the front. This, I hope, will make the area look marginally prettier, like a mini beach perhaps, while I focus on the back garden. This is until I get around to doing something about the cement, which has been poured over (probably) perfectly lovely Victorian tiles. I trudge back through the flat and out again to more and more stones. I sieve builders' rubble from the soil, which I divide into separate piles, to be bagged up for the tip or hidden in pots to be covered with compost and eventually planted up. I advertise stones on Gumtree to anyone who could use them, use the nicest to cover the liner around the pond. Everything is still brown, everything is still muddy, everything is still stones.

Yet all it takes is one little thing to lift me. Like the robin that comes to eat the worms or the *thuuuuur* as a house sparrow flies overhead, still from one garden to the next, on either side of me but nice all the same. The buff ermine moth hiding in the corner – what are you doing here?

Today the day starts like any other. But there's a clear sky and I have the whole day to shift stones. I drink tea and lean against the wall, assessing what yet needs to be done. Late April and still this. The new shed, which I eventually persuaded my friend Clare to help me carry outside, has made a difference to the clutter around the garden, but temperatures are still low and nothing knows what to do with itself. The apple-tree blossom has been trying to come out for about two weeks now. Five little flowers. I'm supposed to take them off, let the tree focus its energy on roots and stems. But I can't. Five little flowers on a stick tree, they're all the garden has. I finish my tea and start raking, gathering stones. There's something in the air, I don't know, an expectancy of something, something I usually feel in March, a change on its way, a shifting of seasons. Or am I imagining it? The five little flowers remain resolutely shut. It doesn't feel like spring yet.

The sun wends its slow way, lowering itself onto the fence, over the reluctant apple-tree blossom, the soil, the stubborn plants. And as it comes the garden changes, the space fills with something, it lifts. And I stop raking and stand in my sea of mud and stones, silently, trying to work out what this is. Something's tickling a memory, something scratching or scraping. A tiny sound, a half-sound. Like the rake but smaller, fainter. It's . . . it's my bee hotel. The one made of broken bits of hollow stems that I reassembled in case I missed a bee and then put out against the south-facing wall because I'm an idiot. I throw my rake down, rush to the far corner where it's propped up against the fence. The scratching is loud here. There's a bee, a big one, in one of the stems. I can hear it but I can't see it, although from faint glimpses of black and the size of the new hole I think she's a hairy-footed flower bee. There's a bee making a nest in this sham of a bee hotel, this sham of a garden. And I'm human again.

I make tea and settle in front of the bee hotel, cross-legged, a child in front of the telly. She's not alone, red mason bees have joined her. I watch them choose holes, arrange themselves, bring mud and pollen. I think they're waking up from the fancy hotel release chamber and flying straight to the stinking mass of familiar stems. I check the other hotels and, sure enough, there are open egg cases. Yet the viewing panels reveal nothing. They love that ugly mass of hollow stems and no mistake; my fault for putting it back out. I return to raking but I can't concentrate, not now. There are bees laying eggs here, making nest cells. Bees laying eggs and making nest cells.

✿❀✿

My friend Helen moves to a three-bedroomed 1930s semi-detached house around the corner. The house is like a time warp. Before Helen it's owned by a woman in her nineties, who bought the house from her mother in the 1950s, who bought it off-plan.

This little old lady had lived here her whole life and when she died her son sold it quickly to be done with it, furniture and all. The house comes with Victorian wardrobes and mirrors, beautifully tiled fireplaces, carpet rails and cleats. In the loft are old newspapers, pianola scrolls, a portrait of a moustachioed man and a Second World War flying suit. The garden is a dream, overgrown and bursting with history and magic. There's a gnarled old apple tree, peonies that have been there for more than fifty years. Ivy has toppled a wall, ancient roses flower at the roof tops. At the back of the garden is a long-forgotten greenhouse, in which I find a hairy-footed flower bee laying eggs in the soil.

But it's a mess, uneven and full of broken glass, dangerous for Helen's two-year-old twins. There's a huge breeze-block garage at one side, which will be knocked down, the roses and apple will be given new life. She calls me: Would you like the greenhouse?

I can't, the garden's too small. I already have a pond and a shed. It would take up another third of the space and look ridiculous. Where would I put it? I need to focus on the mud and the stones, the creation of habitats. I tell her thanks but no. And then I call her back: Actually, can I have it?

You can grow so much more in a greenhouse. You can start plants off, give them a few weeks. You can move the pots and flies and chaos from the table beneath your living-room window and keep them somewhere more appropriate. You can store things that don't fit in the shed. You can take up space that would otherwise be a big muddy mess because you can't afford to buy plants. You can potter in the rain, drink tea or beer and listen to the radio while sowing seeds and transplanting. You can hide here when you want to be invisible. You can be happy.

A bloody great greenhouse, though.

I cycle to Helen's, armed with spanners, a drill and WD40. It's a perfect spring day. The sun is shining and the sky is blue. Everywhere are blackbirds and robins, the hairy-footed flower bee, things flying into this tree or that shrub, startled birds, the low buzz of a bumblebee. Finally, after weeks of cold. It's nice being in a proper garden, nice being among the wildness away from mud and stones.

Helen makes me tea, a laugh in her eyes. She has to take the kids out. She leaves me with my screwdriver and drill, WD40 and the birds. I kiss the children. I'd not taken down, or erected, a greenhouse before. I've no idea what I'm doing. But it's sunny and there's tea.

I start with the window panes. These I release from ancient clips using a screwdriver, prising them from thick cushions of moss. I work steadily, piling them together. The frame is difficult to unscrew, welded together by years of rust. I search online how to shift stubborn screws and watch videos of people drilling holes in them to release the pressure. I follow suit, metal screeching against metal into the clear spring sky. I work around the hairy-footed flower bee, who calmly leaves and returns to her nest as if the house wasn't coming down around her.

It comes down in two days and I drive or carry it in several journeys the half-mile back to mine. It's back up again in a few hours, wonkier than before and with bubble wrap fixed in place of broken panes. An old dog but with plenty of life left yet. I dig the soil and empty old bags of manure into it. I plant tomatoes and aubergines, arrange leggy seedlings on the shelves, pot up others while drinking beer. It's huge and ugly and looks ridiculous. But it's mine. I will grow things in it.

Sheba pants in the heat. She's got her choke-chain on again. Mum says it's cruel but Granny insists on it, says Sheba's a handful. She pulls on the chain, exploring this smell and that. Finds interesting things to wee on. I skip along to one side of them, trying to keep up. We'll let her off when we get to the fields, says Granny.

We're walking along a quiet country lane bordered by recently laid dead hedges made from the pruning or clearing of something or other – branches and saplings. The smell of sawn wood lingers as birds hop through new territory. Sheba explores as far as her reins will go. Her feet paddle in the undergrowth until she's pulled back onto the road. The heat is tempered by dappled shade from trees knitted together at the canopy, each one reaching across to touch the other, like Michelangelo's *Creation of Adam*. I feel I'm walking through a tunnel of trees, that I've been swallowed by trees, their shadows ripple on the tarmac beneath us like water. The gentle rustling of leaves brings a calmness that, aged eight, I've only just started to appreciate.

The road is narrow as roads often are in the country. Two cars couldn't pass here comfortably and there's no pavement for grannies and grandchildren and panting dogs to walk along. We take centre stage where the road rises a little, not along the edge where it crumbles into the ditch. The cars can see us better this way, says Granny. We can move to the side if we need to let them pass. I find the ditch more interesting, though. Sheba does too, I can tell.

There are no cars. And few sounds on this May day except the rustling of leaves and the panting of dog. Birds are dotted about. I go to climb my favourite tree while Granny and Sheba slow down but don't stop. Granny, Granny! Look! She waves and Sheba wags. I love this tree. It's an oak, Mum says. Rooted deep in the ditch, it appears shorter than it is and it has the perfect arrangement of branches to climb up to a sort of platform you can sit in. Must have been coppiced or pollarded at some point, says Granny. I stare blankly from my vantage point. I can't see much from here, just fields beyond the dead hedge on either side, a bit of a farm building but no tractors. I could sit here all day, bring a picnic here, read a book. But it's in a ditch off a narrow country lane. Suddenly I feel silly and climb down. I wish this tree were in your garden, Granny.

My feet slap on tarmac as I run to catch up but Granny turns, suddenly, and stops. Ssssh! She puts a finger to her lips as I creep towards her. What is it, I whisper. Listen, she says. Can you hear it? We stand as silently as we can, Sheba panting and pulling on her lead, leaves rustling in the trees. There, she says, do you hear that? I listen hard. There's something in the distance, like someone blowing on a siren whistle but not quite. *Whoo-hoo. Whoo-hoo. Whoo-hoo.* We stand together in the road as a smile beams across Granny's face. The first of the year, she says, realising it's my first one ever. Come along, dear, she says, let me tell you about the cuckoo.

We turn left into the field and Granny lets Sheba off her lead, who bounces immediately in the direction of the large pond in the distance. Granny tuts. She makes a half-hearted attempt at calling the dog back but gives up, she knows it's futile. Where were we? Oh yes.

The cuckoo is a shady character who flies here from Africa and lays eggs in the nests of other birds, such as the dunnock and reed warbler. It's quite big and grey, with a barred chest. Granny will show me a picture later. Only the male makes the *Whoo-hoo* sound, she says, the female makes a sort of cackle, and I'm not surprised. If I believed any of this, which I don't. It's perfectly natural, she keeps saying, it's just nature. But I'm cross on behalf of reed wobblers and the other one, and I don't think it's very nice. I take it all in, albeit suspiciously, until Granny tells me that the baby cuckoo then pushes the other birds out of the nest so it gets all the food for itself. This is just mean! I ask lots of questions but one I keep returning to is: Why? Because it's nature, says Granny, and we shouldn't impose human values onto wildlife. The cuckoo is not bad, it's just a cuckoo. She tries to tell me about the fox that steals the chicken, the hedgehog that eats the slug. Is that any different? she asks. But there's something about the laying of eggs in another's nest and then flying back to Africa without even seeing your offspring, who then go on to murder their adopted siblings, which is a bit too much to bear on this late-spring jolly with Granny and Sheba and my favourite tree. I don't think I'm being unreasonable here, Granny.

I sulk for a while as we walk along the field edge, me collecting bits of sheep's wool caught on the barbed-wire fence. It feels oily and leaves a film on my hands. Sheba bounds up to us, covered in pondweed, and we laugh. Stupid dog, says Granny, I'll have to bath you later. We come to a stile and then turn back, the pond to our left now and the road ahead of us leading back to the tree and Driftwood and maybe that cuckoo. I hope I don't hear it again. I don't understand why it made Granny so happy. Something about summer being on its way, she said, the end of winter. I can't stop thinking about those chicks being pushed out of the nest. Do they die, Granny? I should think so, she replies, but they won't go to waste, something else will come along and eat them. By the time we come out of the field back onto the road my whole world has shifted. Sometimes it's better to not know things.

I spend my days planting, raking stones and soil. I'm behind schedule (there is no schedule) and I need to get on. The books tell me to raze the whole thing, rake it over, add compost and/or manure and start from scratch on a blank, perfectly prepared canvas. Draw a design, they say, work out what should go where. And then plant. I don't have time for that. To have a garden – any semblance of a garden – I need to act quickly. I can't be waiting for the removal of a million stones and bindweed roots, the levelling and enrichment of soil, the 'plan'. I'll miss summer. I have to work, one scrap of soil at a time. Forget the books, it's May.

I carry the last of the decking planks through the flat and leave them outside to be collected, marking seven months since I unscrewed the first piece. The pond is slowly filling up. Bits and pieces of the back border have long been planted – the apple, teasels and odds such as the common bistort and Japanese anemone. There's white clover and comfrey from Dad's allotment, the rooted honeysuckle cuttings, growing lady's mantle and globe thistle from Mum's garden. A thousand foxgloves that haven't flowered – next year they'll be busy. Everything is tiny, fragile, susceptible to the attentions of slugs and snails, a lack or excess of water, sun, shade or wind, or of being forgotten by me; I'm their guardian and keeper yet also the main benefactor of their success. I have to keep on my toes.

I rake soil, sow seeds, forget about them, sow more or plant things impatiently where germination hasn't been quick enough. Greenhouse shelves burst with maturing plants not yet ready to be planted out, cuttings of box, perennial wallflower, climbing rose, honeysuckle – anything I had or could get hold of is cut and plunged into gritty compost to increase stock and fill space, gifted by Mum, Helen, snatched from the park, the street. My secateurs forever in my bag, stolen seeds and rogue stems spilling from my pocket.

I scatter seeds of love-in-a-mist into gaps in the borders. I weed selectively: mostly avens, herb robert and ivy-leaved toadflax, leaving a bit for the wildlife. Everything in moderation.

The couple four doors down are gardeners. Good gardeners, get-to-work-in-a-cold-spring gardeners. She leaves plants on the pavement for passers-by. One day a

Geranium phaem and some Welsh poppies, another day a hosta so ravaged by snails that I have to cut it back to its rootball and hope it will start again. I wonder if it's a game, if she can see my garden from an upstairs window, if she's leaving things out for me. I walk past, hoping to bump into her, say hello, see if she's left me any more presents. I leave the house to meet friends and then have to run back with an armful of treats. Sorry, there were some plants. A bag of plants.

The greenhouse has left me little room for anything else. But it's good, in a way. A whole room of tomatoes and seedlings taking up space in a tricky part of the garden I might not have got around to planting this year. I can concentrate on the rest of the garden now, the back border and the side bit around the pond. The space in between? For lawn perhaps.

I like a lawn. I like to sit on it, stretch out in summer, delve into the thatch, looking at ants. It sets off the borders nicely. I can grow clover and daisies and dandelions in it, the starlings can forage for leatherjackets in it, foxes can eat peanuts off it. I'm sick of people saying lawns are sterile and bad for wildlife, that they're a monoculture, high maintenance. Better to plant flowers for bees, they say. You can lay turf and in its first year it will be a monoculture. Some Italian species of rye grass that's hard-wearing and suitable for anything from being turned into a football pitch or used as a landing pad beneath a climbing frame. But it's not a monoculture after the dandelions have found it. After the daisies and the plantain have seeded in. After the other grasses, the native species, the fescues and the meadow grasses have reclaimed their space. Mow it weekly into stripes and dress with a weed and feed every autumn and you'll kill all of that, of course. But leave it to write its own rules and it's one of the most diverse habitats in your garden. Really. Some species of ground-nesting bee need closely cropped lawns. Green woodpeckers use short lawns to find ants, hedgehogs seem to prefer foraging on short grass. Let it grow a bit longer and insects will hide among the blades, house sparrows will pick them out and feed them to their young. Let the 'weeds' flower and seed, let caterpillars hunker down in the thatch. Watch the wildlife come in. A mixture of heights and weeds makes the perfect lawn, in my book. Stripes are overrated.

And yet all I have for this is a pathetic shady strip between pond and greenhouse and my hidey hole at the back, my little space between the pond and the climbing rose, that gets the most sun, where I can squeeze in my deckchair for the rare occasions when I sit down. This, reluctantly, is to be my lawn, my 'seating area', where I will lie and read books, delve down and look for ants. I dig the earth over, remove stones, rake it level. Water and sow seed. This garden will yet be a garden.

Summer

My garden was grey and now it's green. Sort of. I feel like a failure, though, a charlatan. Nothing looks done. Nothing is beautiful. But there's been a change, a shift I can't put my finger on.

When I work now I can feel the house sparrows watching me from the holly and buddleia on either side of the garden. I think they've started to come in but I can't tell. The only window looking out is from the back door into the gully, and it's frosted. And they're so shy when I'm out here with them. Sometimes I hear a *thuuuuur* as one lands on the wall above the pond, cranes its head, spots me, *thuuuuurs* off. Are they bathing in the pond, I wonder. Eating, what? They ignore the hanging feeders of sunflower seeds. I empty them into the ground feeder and replace them with mixed seed. Just a little, in case they don't like it. One outside my bedroom window and the other at the back, with 'Shropshire Lass'. Maybe that will bring them in.

There are other birds too. Heavy birds, I think. Stones shifted around the pond, white tail feathers, big green shits. I look up at cackling herring gulls on the roofs.

I have a greenhouse and soil that will be a lawn. Plants that will grow to fill the space. Eventually. Maybe. The plants, still small, sit politely where I planted them, not yet sprawling as they should do in a few weeks. June and still so much mud.

✿❀✿

She sticks her bum in the air to tell the boys she has mated. Bright orange, it is, orange for Leave Me Alone. It's not a bum, really, but a scopa on the underside of her abdomen, a patch of hairs or a 'brush', used to collect pollen. And with it held high, like Mary Poppins in full bustle, she flies, unmolested, from one flower to another, gathering food to feed her young.

She visits drumstick alliums and sweet peas, ornamental thistles and perennial wallflower, she's not fussy. She dives down into a thistle head and all you see is wiggling orange bum as she swims across its anthers. She takes deep drinks of nectar. She doesn't rest, launches herself into the air again, now back to her nest where she regurgitates the nectar and brushes the pollen off her scopa. Mixes them together. She backs out of the nest and then backs into it, lays an egg. Then she returns to the garden. She's looking for something else now. She flies around a bit, lands on a rose leaf. She clasps the leaf between her legs and chews into it, working her way around it as a pair of scissors. She takes seconds to do this, cutting and rolling as she goes, the perfect elliptical disc. Heavy now, her wings have to work harder. She lifts off like a helicopter, *brrrrrrr*, the disc of leaf rolled up between her legs, the weight of it pulling her down before she gains enough momentum to lift herself skyward again. She carries it to her nest and fumbles with it, unrolls it and pushes it in. She makes it wet with something like spit, wallpapers it to the sides. She pastes it into the corners. She's locking her baby in. Locking her egg with its parcel of pollen mixed with nectar, into its little leafy hollow. She works in a circle, sealing the leaf to the wall so nothing can get in. She inspects her work thoroughly. And then she backs out again, chews a piece of leaf again, flies back again and begins building again. In front of the egg with its pollen and nectar, in front of the leafy hollow, she starts making another cell, another little nest for another little babe. The first section of cylinder is made from four leaves pasted together, like a closed daisy capped off two-thirds of the way along its petals, in which the first egg lies. Now she pastes four more leaves to the existing ones, lengthening the daisy. Sometimes she ignores 'Frances' and takes a disc from an evening primrose leaf or even its bloom. The new nest cell is prettier than the last, yellow and green. She returns now to the drumstick alliums and the ornamental thistles, wiggles her abdomen to gather pollen, fills her belly with nectar. She flies back to the nest, deposits her load, flies again to the flowers for more. A few trips now. Then, when the nest is ready, she backs in and lays her second egg. Flies out to get a leaf, seals her second baby in, pastes the second leaf cell down. She gathers more leaves and pretty pink petals, extends the cylinder further, makes the third section and lays the third egg. Cut leaves, gather pollen, lay egg, repeat. It takes all day and she's barely started. It takes all day to lay three eggs.

I wait until dusk before I sneak a peek in the bee hotel. It's not nice to disturb them during the day and they can abandon the nest if they feel unsafe. I peel back the viewing panel and see her cylinder of leaves, beautifully arranged in a variety of colours. She's resting in the newest one, pops her head out to see what's going on. It's just me, little leafcutter bee. I'm just seeing how you're doing. She shrinks back into her cylinder and I gently close the door on her, return her to darkness.

There's no way of knowing where she came from. A bee hotel in someone else's garden or an undisturbed cavity in a wall or tree. Only red and blue mason bees nested with me until now, this leafcutter is a pioneer. I'm so happy. The rose, 'Frances E. Lester', is barely six months in the soil, the thistles and alliums their flowering first. Yet here she is, oblivious to the bareness and the smallness, the ungerminated grass seed, the expanse of stones. There's pollen and nectar, the right type of rose leaf, a bee hotel to nest in. She's here, only a few months after the garden was released from its prison of decking. My first leafcutter bee in this half-made mess. My heart swells with pride.

A queen white-tailed bumblebee founded her nest in a hole a mouse had tunnelled into the border, beneath the ice plant, and long since abandoned. The ice plant is ancient, fifty years perhaps, planted by the old lady who lived in the house her mother bought off-plan in the 1930s. The bees were safe here, deep below the surface among gnarled roots and well-formed walls, the roots holding the tunnel in place. Smelling sufficiently of mouse to ward off predators, the queen conducted her business unseen and unmolested. Only her workers braved the outside world, foraging from flowers above ground to bring nectar and pollen to her babies. The nest grew and grew: more babies that became workers that foraged food to feed the next lot of babies. All beneath the ice plant, the ground, all hidden from view. Bumblebees nesting in the border. Who knew?

Helen.

It was early May when she told me she'd seen bees going in and out of that hole, that she thought there was a nest. But after sitting in front of it for half an hour, I confidently told her she was mistaken. One had probably been hibernating there, I reasoned, there's nothing going on now. So she got on with her job of razing the garden. She had the garage knocked down, levelled off the lawn, took cuttings of plants she wanted to save and dug up and threw the rest into a skip (which I later rescued and took for my own garden). She dug up the ice plant because she wanted to keep a bit of it, trim it down, resurrect it, but in doing so she dug up the bumblebee nest. She turned the safety of that old mouse hole into a death trap. And she didn't know what to do, she didn't know how to fix it, so she called me.

I cycle over while she fetches two jam jars and a bit of cardboard from the recycling bin. In digging she's filled in the entrance hole and now five white-tailed bumblebees, legs loaded with purple and orange pollen, angrily buzz around it. The ice plant, half-dug out of the ground, hides the rest of them.

I learned to catch bumblebees in jam jars from F.W.L. Sladen who, in 1912 at the age of just twenty-six, wrote my favourite book: *The Humble-bee*. A bumblebee fanatic, he would trawl the Kent countryside in his horse and trap looking for nests, which he would dig up and take home with him. In 1911 he dug up more than a hundred. Some he kept in the garden and others he would raise in the parlour. I like him, he makes me want to keep bumblebees in my bedroom. In *The Humble-bee* he describes the lives and habits of bumblebees, explains how to catch and rear them, documents his observations of bumblebee nests. More than a hundred years later he's still the leading expert on various facets of bumblebee behaviour. He's my bumblebee hero.

'As regards the taking of the nest,' he writes, 'I have never been able to improve on a method I devised when a boy. The apparatus needed is quite simple – a strong trowel, two glass jars with narrow necks, and two squares of card large enough to cover the mouths of the jars; also a little box to hold the comb, though in my boyhood I used to carry it in my pocket handkerchief.'

Using his trowel he would dig up the nest and catch errant workers in the jam jars. 'This jar I stand in a convenient place on level ground about a yard away and place a stone or lump of earth on the cover to prevent the wind blowing it off. Continuing to dig out the hole, it is not long before I'm greeted by another worker from the nest; she is promptly captured in the other jar; this jar is then placed, mouth downwards, on top of the first jar, the two cards drawn out and the two jars given a vigorous shake; this causes the bee in the upper jar to drop into the lower jar and the stone replaced on it. Thus all the bees that come out are caught, one by one, in one of the jars, and collected in the other jar.'

Today's bumblebee experts have since told me that a spade is a much quicker and more efficient method of bumblebee removal – you simply dig the nest out of the ground and lower the whole thing, bees and all, into a shoe box or similar lined with grass clippings and moss. But I always start with Sladen, if only to sit for a moment and pretend I'm in the early-twentieth-century Kent countryside catching bumblebees from the hedgerows while a warm, snorting horse waits for me with my trap.

Helen is frantic: worried she's killed the bees but also allergic to their stings and concerned her two-year-old twins might be as well. I wear a vest and shorts, drink tea, play with the twins. I'm too relaxed, she says, to be catching bumblebees. I seize the five workers in a jar and look for the entrance hole beneath the sedum. It's completely blocked, and has been, Helen admits, since the previous afternoon. I worry for the bees. The five outside would soon lose energy and die – no huge loss to the colony – but those inside it are in danger. I think of Chilean miners trapped in the bowels of the Atacama Desert. How long can a bumblebee hold on without food, water, light and air? There's no choice but to dig the sedum out completely and see what's beneath it.

I get a fork while Helen and the twins stand back. I don't know what to expect; silence would be the worst thing, signifying a dead nest. A queen and a few workers, suffocated in the earth. Or a crushed nest, destroyed from the weight of the earth piled onto it. No; I want them angry and alive.

I get them. A cloud of furious bumbles erupts around us. There are hundreds of them, the largest nest I've seen, and in early June, after a late start to the year, this is a good sign. The fact they are mad means they're alive and kicking. And the fact they are swarming around my head means I need to act fast.

I lift the sedum clean off the nest, gently popping it on the ground nearby, and the bees instinctively fly to it, looking for their entrance hole. The nest – a small mound of grass, presumably assembled by the long-gone mouse – is miraculously still intact. I can just see the waxy pots poking through, in which bee grubs and food – a watery mix of nectar and pollen similar to honey – are stored. From deep within I can hear the fierce, high-pitched buzzing of the queen. As I dash around dodging bees, in my shorts and vest, Helen tells me I look happy. It's one of my favourite afternoons of the year.

Sladen's method long forgotten, I retrieve a shoe box from Helen's recycling bin, fetch a spade and gently lift the nest – bees and all – and lower it into the box. Most workers settle down but there are a few stragglers still focusing on the site of the original nest that need showing the way. I collect them and drop them in their new home and, as soon as they land, they stop buzzing. Home at last. I reunite the five originally caught in the jam jar, drink more tea and eat a chocolate éclair, and cycle home again.

I first moved a bumblebee nest ten years ago. It was before my ex, Jules, and I had moved in together; she was living in a rented flat in Manchester, and her flatmate Jonny had thrown a mouldy duvet out into the back yard. I don't know what he expected would happen to the duvet. Most people would just wash it if it was mouldy or dispose of it, but it was in those post-student days of not quite being a responsible adult and so Jonny, who wasn't very house-trained at the best of times, chucked it where he couldn't smell it.

But a bumblebee smelled it. A red-tailed queen. She will have found the duvet in March or April, laid six eggs and brooded them like a bird, while she survived on a small portion of pollen and nectar she had gathered for herself. When the eggs hatched into grubs she will have left the duvet and gone out to gather pollen and nectar to feed them, and then, after they had grown and metamorphosed into worker bees, they will have gone out to forage for nectar and pollen while she brooded another six eggs. Gradually the nest will have grown, until the neighbours noticed buzzing and complained to the landlord. He phoned Jules and said, Move that nest or I will. And she said, What nest? Neither she nor Jonny had noticed the bees.

It was July 2006. Bumblebee Conservation Trust had just launched and had managed to secure a front-page story in the *Independent*, announcing that bumblebees were in trouble. I remember clearly which shop I had seen the paper in, next to the takeaway on the Oxford Road, near the swimming baths. Bumblebees piled high on the floor in front of the sweets. I found Bumblebee Conservation Trust online and fired off a message, detailing the nest in the duvet and the landlord and the neighbours. Ben Darvill, now a friend, wrote back, telling me what to do: wait until darkness because bees don't fly at night and therefore can't sting you, cut the nest out of the duvet and gently pop it into a box lined with grass and moss with two taped-up entrance holes, and drive this to a nearby park or allotment, ideally eight miles away because the bees won't try to return home. Untape the entrance holes the following morning.

It seemed so easy. We called it Operation Bumblebee. Jules made a beekeeping outfit using old net curtains but I wore jeans and a T-shirt, believing Ben when he told me bees don't fly at night so they can't sting you. They can if you shine a torch at them, Ben. I was stung twice but it didn't really hurt.

We cut out the nest and moved it into the shoebox. I'm not sure how easily the nest went in. I imagine a lot of honey pots were spilled, bees lost. But we managed it, we moved the nest and drove it to my allotment. The next morning I cycled there and removed the tape covering the holes, just as Ben said to do. I went to work but I was late and I couldn't really concentrate. After work I went back to my allotment to check on the bees. And that was it for the rest of summer: my new-found love affair with bumblebees, something inside me unlocked, a new version of me found in a mouldy old duvet in a back yard in Old Trafford.

As often as I could I would sit and watch them. I'd see workers come in and out with legs laden with pollen, males coming and going. I saw a lot of dead bumblebees, which made me worry I'd somehow contributed to their demise, and I watched a daughter queen dig herself into the ground, just next to the nest, which initially I thought was bumblebee suicide. I bought books on bumblebees and learned that red-tailed bees have a preference for yellow flowers, and seem to like nesting in damp places, like walls in moorland and, perhaps, mouldy old duvets thrown out into a yard. God knows what would have become of her, or me, if it hadn't been for Jonny.

I've since moved three bumblebee nests, only ever because they have already been partially dug up due to a garden redesign or similar. That first nest was going to be destroyed by the landlord. The second was found at the base of a compost heap that had already been moved. The third was under a bamboo hedge that was being grubbed out. I gathered the bees and their broken home into a box, which I placed just a couple of metres from the original, so the bees could carry on as before. One I took home with me, which I wasn't allowed to keep in the bedroom and document the bees' behaviour, as Sladen did all those years ago. I don't move nests willy-nilly. We must, unless absolutely impossible, leave bees to get on with being bees. We make life hard enough for them as it is.

Helen's garden, now a field of raked earth, is no place for bumblebees. But you can't relocate a nest during the day, despite Sladen's advice to do so in the afternoon. There are too many workers out foraging for food. I leave the boxed nest of bees open, exactly where the sedum has been, for the bees to get used to their new home, settle down and continue foraging for nectar and pollen. At 10 p.m. (after sunset) I text Helen to remind her to put the lid on the box, which she dutifully does. I get up at 4 a.m. the following day, drive round and pick them up. There are no bees circling the nest, none missed the 10 p.m. curfew – it's complete.

The bees are angry as I drive them to mine. They buzz loudly and I can hear the queen above the din. But within five minutes we have arrived. I place the bees gently at the back of the garden, remove the tape from the entrance holes, and go back to bed.

✿❀✿

The white-letter hairstreak is a shy little butterfly. Keeps to itself. Doesn't make a big song and dance like some of your other butterflies. Doesn't parade itself on buddleia like your peacock or your red admiral, all strong and bold and brightly coloured. The white-letter hairstreak is brown and small. When it rests you can see its orange wing edges and the delicate letter 'W' written in white on its underwing. Little, insignificant tail streamers. Except you won't because it lives at the top of elm trees and you probably haven't noticed it.

It doesn't get out much. A bit like the house sparrow, it spends most of its life with a handful of others in the same place – in the canopy of one or two elm trees. Always and only elms. It starts its life as an egg on a twig at the outer edge of the tree. Endures winter like this, exposed to the vagaries of hungry birds and tidy councils who like to trim trees into a lollipop shape – butterfly eggs in the bin, a life cut short before it's started. Those that survive hatch into a caterpillar in spring, eat flower buds before pupating and emerging as an adult in June. It flies until the beginning of August, mostly around the canopy of the tree it started life in, feeding on the honeydew secreted on its leaves by aphids, mating and laying eggs on its twigs. Sometimes it comes down to feed on bramble and thistle nectar. Sometimes. It's most active in the morning, apparently. If you want to see it you need to find an old elm tree and take your binoculars. Look for a triangular-shaped, chocolate-coloured insect bobbing jerkily in the tree top. You might be lucky.

This shy, retiring butterfly used to be fairly common but Dutch elm disease all but wiped out its caterpillar food plant, its winter hibernaculum, its home. It can't just go and live on an ash or an oak, it's evolved to live and feed on elm, it is entwined with elm, relies entirely on elm. So, along with the elm, it is all but wiped out and no one has noticed.

I've seen it once. Brighton and Hove has the National Collection of elm trees. Something to do with the Downs forming a natural barrier to the beetle that spreads Dutch elm disease, and something to do with the conservationists of Brighton and Hove being prepared by the time it arrived here. There's an army of volunteers who check their local elm trees for signs of damage. At the first hint of disease the tree is felled, sacrificed to save the others. Most of Brighton and Hove's street trees are elms. You can tell because they flower before the leaves develop, streets of red lollipops in March and early April. And when the leaves come and the seeds disperse, delicate little samaras, winged seeds in a papery tissue, become lodged in the gutters and drains and pavement cracks of the city. And no one notices.

I interviewed the local white-letter hairstreak expert a couple of years ago for an article I didn't end up writing. She took me into the hills, to Hollingdean Park where there is a stand or two of mature elms. It was early August, late in the white-letter hairstreak season. We scoured the tree tops with binoculars, I couldn't see anything, but with her trained eye she pointed at a fluttery thing against the blue sky. I squinted and caught the briefest glimpse before it disappeared into the canopy – my first white-letter hairstreak.

White-letter hairstreaks like mature elms. Not the newly planted hybrids in the city centre. The older the better, where annual generations of butterflies have lived for many years. There are some on the Level and in the grounds of the Royal Pavilion. And there's the Preston Twins, the two oldest elm trees in the world, situated by the north-western entrance to the park on London Road.

There's a good chance of seeing white-letter hairstreaks above the Preston Twins. A good chance of seeing some of the most threatened butterflies in Britain in the canopies of some of the oldest trees in the world. Across the road from the Rockery, where I run and gawp at frogspawn in spring, are these two trees and their shy brown butterflies, four hundred years of Brighton and Hove's history neatly packaged into a few square metres.

Summer now, shorts and a vest, no cumbersome hat and gloves but a small bag packed with binoculars. I head out into Sunday morning, less of the bustle and traffic of the weekday. I run up past the closed chippy and Hove station, past the Tesco Express, into the hills to the Droveway, the route I last ran to gawp at tadpoles. Early July, the vest was, perhaps, a little ambitious, and I run faster to warm up. Down South Road and across Preston Road, I head towards the north-western entrance to the park.

I stop, breathless and sweaty, and lean against railings on the edge of the road. Look at them, these great old things, two old dears guarding the city since 1613.

'English' elms. Yet they were brought to the UK by the Romans. They're huge, with thick trunks from which a hundred tributary branches fan out like veins, like water. I fish out my binoculars and scan the canopy. A slight wind has picked up and the branches and leaves are quivering. No white-letter hairstreaks yet. I lower the binoculars to look into the trees, into an ancient world hidden from passers-by on the road. Far up, a baby jay sits, unmoving, on a branch. Elsewhere a charm of goldfinches twitter. If I searched closer I'd find ladybirds and other beetles beneath the bark, ants, leafminers, grubs of things I couldn't even begin to identify. Each tree a city, housing the tiniest insects up to birds and small mammals – mice and squirrels, bats perhaps. And, somewhere, the white-letter hairstreak butterfly.

I scan the canopy again, take the binoculars around the edge of each tree, the outermost branches where tits forage. White-letter hairstreaks like to hide in the canopy. The males come out to fight in great aerial battles; both sexes, occasionally, come down to feed. But mostly they stay hidden. Mostly, you have to be patient.

The sun disappears behind a cloud and everything becomes quiet, except the road behind me, where traffic is picking up. I try to block it out. I scan the bark with my binoculars, look inside the great hole Mother Nature carved out in one of them. These trees with their four hundred years of gnarling, four hundred years of twisted growth and of being home and food to thousands, being part of the landscape for a growing town, a new city. Tell me your life, Preston Twins. Part of the old Preston Manor and not far from the Droveway, these trees would have seen the comings and goings of farmers, the felling of the Rookery to create the Rockery, the developments of crown estate into parkland, of the road, the first buses and cars. These trees and the butterflies, too: each generation, each year, a little different from the last.

The sun comes out and I tune my binoculars back to the canopy. Nothing but the gentle waving of stems in the breeze. I keep looking, keep scanning. I shiver as my body cools down from running; I should have packed a sweater.

There. A flicker of something. A butterfly. Flying out and then in again. Is it feeding? I can't tell. I use my binoculars to zone in, focus on it. It rests on a leaf, a little chocolate triangle, a white-letter hairstreak. Just a hint of it, a glimpse, and then it's gone again. But definitely there. I continue looking but see nothing else. Is that it? Again? Two white-letter hairstreaks glimpsed in a flash? I like them more for that, somehow, shy and elusive as they are. As well as being cold, my eyes hurt and my neck is strained from looking upwards. But I've done it. I've seen a white-letter hairstreak butterfly in one of the oldest elm trees in the world. I pack my binoculars back into my bag and run back home to the warming day, a little more in love with these butterflies, a little more in love with Brighton.

✿❀✿

Helen gives me spare turf from the new lawn she's laying in her own garden. I lay it around the greenhouse and in my hidey hole at the back. It makes the garden look instantly green, pulls it together somehow. The greenhouse is bursting with ripening tomatoes, chillies and aubergines, the red-onion squash is threatening to take over. The teasels are nearly finished and the *Verbena bonariensis* and *Knautia macedonia* are gearing up to take centre stage.

The house sparrows are finally here. Ever since I changed the seed in the feeder. Now when I wake I hear them outside my bedroom window. When I open the back door there's a flush and a flurry of action, little brown bodies flying up into the holly to the left, the buddleia to the right. They launch from the pond leaving droplets in their wake. They're settled now and I love them for it. Eventually they'll have shelter to rival the holly and buddleia, long grass to gather caterpillars for their young. I made a home for house sparrows and the house sparrows have come.

When I work they are with me. Never in the garden proper, although sometimes four or five come and sit on the wall, crane their necks, daring each other to hop in with me still around. They never do. It's like they're telling me, I'd like a bath now, I'd like some food now. Go back inside and let us have the garden again. Cheeping alarm calls to each other while flying, impatiently, between the holly and buddleia.

In the morning, sometimes, I open the curtains slightly and sit in bed with tea, watching them through the crack. They're still so shy, fly off at any sudden movement. But if I'm still I can watch them for a while. They arrive in a big gang of around twenty. They use the box bush for shelter, some keep sentry on the wall. There doesn't seem to be a pecking order; if one wants to eat and another is in its way then it simply lands on the other's head. Maybe more dominant birds land on less dominant ones or maybe it's just a free-for-all.

It's getting warmer and the garden is growing. Borage is leafing up and developing flower buds, roses are weaving into their space. I find plume moths amongst the foliage, hoverfly pupae on leaves, froghoppers, a centurion fly. Life.

✿❀✿

It's 8 a.m. when I first step outside. I shoo two large white butterflies out of the greenhouse, where they're forbidden from laying eggs on my broccoli and Italian kale. I walk the few steps around the garden and pause to watch them mating before the female drops down onto a nasturtium leaf and rests awhile. Before long she starts laying. Sitting on the edge of the leaf she curls her abdomen underneath it and squeezes out an egg on the underside in a single motion: *dab*. She curls her abdomen back up for a second and then reaches it beneath her again: *dab*. A few more times: *dab dab dab*. She can lay as many eggs as she likes on my nasturtiums, that's what they're there for. Others have found them before her and the whole back fence is a mass of nasturtiums laden with eggs and tiny caterpillars. It's lovely to see (as long as they stay away from my broccoli and Italian kale).

Elsewhere is a butterfly desert. On runs now I look for caterpillars. I scour nettle beds in Hove Park and St Ann's Well Gardens, along Brighton's train tracks and in forgotten beds in supermarket car parks. Caterpillars mark summer. They tell us the world is still turning, the systems are still working. But this year I can't find any.

Many of Britain's fifty-odd butterfly species lead obscure lives in carefully managed habitats but there are around twenty that would come into our gardens if we'd let them. Butterflies need nectar to give them energy to fly but they also need caterpillar food plants: leaves on which they lay eggs and which their caterpillars can eat; leaves that enable them to complete their lifecycle. Each butterfly has its own food plant which it has evolved alongside: the small and large whites so hated by kitchen-gardeners lay eggs on brassicas: cabbages, broccoli, kale. But they will also lay on nasturtiums and that's why I grow them. The holly blue lays eggs on holly and ivy; the brimstone on common and sea buckthorn. The most colourful ones that we draw as children, that bounce around on buddleia in late summer, all lay eggs on nettles.

When I first started searching for caterpillars, I didn't know what to look for. I would stand in front of a patch of nettles and feel lost. I didn't know whether to look at the tips or further down. I didn't know what the tent constructions the caterpillars made for themselves looked like, whether they resembled cobwebs or if the things I found on nettles actually were cobwebs. All I knew was that small tortoiseshell, peacock, red admiral, comma and painted lady butterflies laid eggs on nettles and that the caterpillars of some species fed communally and sheltered beneath large silk tents, which they moved every so often as the nettles were eaten. I was told those of small tortoiseshell and peacock were easiest to find.

It took a few years of looking but one day I found them. I was in Cornwall with Mum, Ellie and my little sister Anna, who was born when Ellie and I were teenagers. It was the summer everything changed, the summer of my break-up. The week was marked by giant blue skies and running to the sound of skylarks, and a backdrop of great, crashing grief. We drank Cornish ale and took long walks. I found a slow worm. It was lying, cold and stiff, in the road but still, just, alive. The day was cold and full of rain. I picked it up and popped it in my pocket – a warm place for this cold-blooded lizard. We headed to the nearest pub where we knew there was a large compost heap, all warm and with plenty of food – a happier place than a cold, rain-soaked road.

By the time we got there it had died. I was devastated. It had probably been run over, it probably couldn't have been saved. Mum choked back tears as I laid it down anyway, in case it would revive, and covered it in a thin layer of grass clippings for warmth. At the edge of the compost heap was a small patch of nettles. And on these nettles were lots of little black caterpillars. Oh.

With an old ice-cream carton I returned later to the compost heap, and snipped nettle leaves laden with caterpillars into it. I took only six; Mum named them after the wives of Henry the Eighth. Each day I fed them fresh nettles and cleaned out the box. I watched them shed skins as they grew and developed their intricate small tortoiseshell markings. I videoed them munch leaves in an anticlockwise direction, spin into a chrysalis. Poor Jane Seymour was parasitised by something gruesome.

It's funny, writing about that time. That week was horrendous. Despite the skylarks and the walks, the soul-searching and the comfort from those who are a part of me, I was desperate, in shock, a raging, crying, running mess. Each day I ran along the cliff edge until I was breathless and felt sick. And then I would return to my family and we'd get ready to take another walk, have another lunch. Because all we could do was carry on. And I know now that, despite the running and the first, enormous wrench of shock and loss, I was OK, I was functioning. I was running, I was eating, I was trying to rescue slow worms. And I was keeping small tortoiseshell caterpillars as pets.

I learned a lot about caterpillars that year. I've raised more since, once a whole load from a patch of nettles that had recently been sprayed. It's a nice way to measure summer, to involve yourself in the delicate cycle of life, remind you that the world still turns no matter what is going on around you or within you. And it's easy, really.

But this year there are no caterpillars. Not on the nettle beds in Hove Park and St Ann's Well Gardens, along Brighton's train tracks or in forgotten beds in supermarket car parks. The nettles are eerily quiet, eerily lush without wriggling things to eat them. And it's terrifying. The smaller things, the systems that so often go unnoticed, are dying without us even realising. Conservationists talk of abundance and percentages and our eyes glaze over but the evidence is plain for all who will open their eyes: butterflies are disappearing and they're doing so on our watch.

Two days after I took those first six caterpillars I returned to the patch of nettles to find it had been strimmed and added to the giant heap. It was too late to pick through the remains looking for caterpillars, they had all dispersed and probably died. Turns out those six I had taken I had saved – are their descendants still flying or did they too lay eggs on nettles that were cut down?

It seems there are more white caterpillars than usual this year but I wonder if they just seem more abundant in the context of the loss of the others. Sometimes I fear large and small whites are the future of Britain's butterflies, that they will be all we have left, and people will still hate them. Garden butterflies need gardens, yes. But in those gardens they need nectar and nettles – specifically nettles grown in sunshine. And that's the problem: those who love their gardens usually baulk at the thought of growing nettles in full sun; those who don't love their gardens still wouldn't want them. It's the same on our verges and allotments: health-and-safety fears, years of castigating them as weeds. Oh, but they're ugly. They'll sting my kids. Add to that the changing role of gardens: they're becoming smaller, more like outside rooms. There's more decking and paving, fewer plants and less soil. And the plants we do grow for them are often fresh from the garden centre and therefore laden with a cocktail of pesticides and fungicides. The butterflies drink it down in the nectar and then what happens? Do we know? No. Can we guess? Yes. We're poisoning the plants and paving over the land and our butterflies are in free-fall. Most of us are too busy and frazzled to care.

I find eggs on my broccoli and Italian kale and transport them, diligently, to the nasturtiums. Already large chunks have been taken from the leaves. The next generation of small and large white butterflies is on the march.

In my hidey hole no one can see me. I'm mostly in the shade, only my feet in baking sunshine. The air is thick with insects; hoverflies buzz around fennel flowers, butterflies tumble among the browning teasels, wasps hunt caterpillars like sharks. Helen's white-tailed bumblebees, still in their box at the back of the border, feast on the nectar from a hundred blooms. It feels like a garden; it is a garden.

The borders are full; there's no space left to plant things. And yet there are plants waiting to be set free from pots and greenhouse shelves. I have nothing to do now except wait. Wait for things to die down, for the tomatoes to ripen, for herbaceous plants to seed and go over so I can cut back, chop, make room for other things. It will be only a few weeks. After that the half-life of winter will force me indoors, to plan and sulk and wish for spring.

The grass is growing and prospecting queen ants furrow among the blades, looking for a suitable spot to start a nest. One couple lands clumsily on the sweet peas, locked in coitus, the male quickly discarded. I close my eyes and listen to the hum of insects. I could be anywhere, I could be in one of the most established gardens in the country. But I'm here. Here in my space with my plants and my bees and my blades of grass and my prospecting queen ants. Beside me lamb's ears and willowherb buzz with a thousand bees. There's a small, rusty-looking bee that resembles the wool carder and it enjoys the lamb's ears with a similar aggressive, darting flight. *Anthophora furcata*, or the toothed flower bee, I think. A new species for the garden. To my right common darter dragonflies lay eggs in the pond. It's only three months old and it's not even full. The red male and greenish female mate on the wall, a marvellous feat of heart-shaped gymnastics known as a mating wheel, which is also odd and uncomfortable-looking. Afterwards the female flies over the pond and pauses, repeatedly, hovering over the water and dabbing her abdomen into it. Baby dragonflies. *If you believe, they will come.*

The house sparrows have seen me and are shy. They hide in next door's buddleia. A few at a time land on the wall and crane their necks cautiously, looking around, before the bravest sails in to take a few sips of pond water. Any movement or sound from me triggers its departure, *thuuuuur*, a wet dog shaking the sea off its coat, a deck of cards being flicked before a game.

They're braver now. Time was I would never see them feed in the garden, let alone use the pond. Now I can sit, albeit in my little corner, and watch them bathe and drink in the pond, or squabble ten at a time over the hanging feeder in the gully.

I worry about them, specifically about next door's buddleia. The tenants have been evicted and there are men in overalls decorating in the flat. I worry someone will come along and chop the buddleia down, leaving nowhere for the house sparrows to be. The house sparrows love that buddleia. When they're in it I can hear them but not see them – that's what they like, to be heard but not seen.

The house sparrow is a sedentary species. It lives in extended family groups in small territories, never straying far from its boundaries. When the food and shelter in that territory is removed, these family groups die out, rather than move on. It can be years before anyone really notices, such are their great numbers, but one day you look out of your window and the hundred house sparrows gobbling all your mixed seed becomes thirty, then twenty, then ten. Habitats are being compromised or lost in our towns and cities – a hedge grubbed out here, a front garden paved there, a 'garden-office' built somewhere else. Green space is disappearing from London at a rate of twenty-five football pitches per year and the house sparrow is going with it.

House sparrows have simple habits: they take shelter in large shrubby plants in which they can be invisible; they're seed eaters but feed insects to their young; they live and nest in loose colonies, often in the eaves and roof cavities of houses. That's it. To survive they need large hedges or shrubs, long grass and unclipped and untended patches of land (better for insects and therefore their predators), and somewhere to nest. But you rarely find these habitats in urban gardens, especially small ones, like mine and next door's. The house sparrow is suffering for our modern tidy ways, for our penchants for a second or third car, for working from home, for not liking birds in the roof, for the lack of time or inclination to have outside space, for our population growth, for the halving of gardens when a house is divided into flats. In rural areas they suffer different types of habitat loss, plus greater levels of pesticides. We're changing more quickly than they can.

Despite its popularity among pollinators for nectar in summer, buddleia isn't such an insect magnet really. Its leaves are unused by most moths, leafminers, flies and aphids, whose larvae the house sparrows rely on to feed their chicks. The thing about buddleia is its size and sprawling nature, and its knack of growing in otherwise barren urban areas: cracks in walls, chimneys and other out-of-reach places. Introduced from China by the Victorians, it has caused problems in the wild because it out-competes native plants that provide much better overall habitats for insects (and therefore, ironically, house sparrows), but it's a welcome refuge in cities. The Regents Canal in London, near where I used to live, is all gas stations, expensive apartment blocks, hipsters and gentrification. But where there are cracks in walls there is buddleia, and therefore still, but who knows for how long, house sparrows. I would run along the towpath and the sprawling mass of buddleia on the other side of the water would be *noisy*. All you could hear, for a huge stretch of towpath, was the *cheep cheep cheep* of house sparrows chatting to each other, completely unseen but heard all around.

Here, in Hove, in next door's garden, the buddleia is one of a few plants along a long stretch of paved-over and suppressed gardens that provides a habitat the house sparrows can use. It's not the best habitat but it's the only one they've got. If it's cut down, they will no longer have shelter when the cats or herring gulls come for them. And so they might stop coming to my garden, and then they might starve or be more exposed to cat strikes or have fewer successful nesting attempts due to fear and stress. The house sparrows were the first species I noticed when I saw this garden and they were the first species I tried to help, but my intention was to increase their habitat, not mitigate against future losses. As I took the decking up from my garden, someone else decked theirs three doors down. As I planted, so others removed. Whether the house sparrows can keep up with all the change I don't know.

Sometimes I run up to Seven Dials and then down to the sea. On one road behind fast-food outlets and opposite a block of flats beneath a rooftop car park, there are huge Regency houses, converted into flats, and a corresponding sequence of small front gardens where house sparrows live. But the birds here are living on borrowed time. Already there are so few of them and they seem to spend most of their time in just one garden. It's the messiest on the street, owned probably by someone elderly and infirm or a buy-to-let landlord who doesn't care, and which the neighbours probably complain about. It has unclipped hedges and long grass and gone-to-seed plants. To some it might be an eyesore, a blot on the landscape of otherwise neatly paved 'gardens' with the odd pot of wind-and-sea resistant geraniums. To the house sparrows it's a lifeline.

When the buddleia is chopped down I hope there's enough for the house sparrows in my garden. I hope there's more than a whip of an apple tree with three branches, twiggy shrubs and immature climbers. In theory, the plants I'm growing are better for house sparrows than buddleia. There's honeysuckle, apple, guelder rose and spindle, two climbing roses and a couple of fluffy seed-bearing clematis. When lush and mature, they'll provide the perfect habitats to nest and take shelter in, and all of them attract insects the house sparrows can feed to their young. But that could take five years, and the buddleia could be chopped tomorrow.

These little brown birds, which make so much *noise*, and steal food from our plates when we're on holiday, are declining at a horrific rate. They're a long way from dying out – still our most common garden bird – but the sudden and sharp drop in numbers should be an alarm call to us all. They're doomed if we don't help them.

Despite all the lushness and the ripening apples and the huge winter squash and the prospecting queen ants and egg-laying dragonflies, I'm worried about my house sparrows. Despite everything I've achieved in my garden, it will never be enough.

The great yellow bumblebee carves out a niche existence in the wilds of the Highlands and Islands of Scotland. It used to exist all over Britain but has declined by 80 per cent in the last century, squeezed north due to habitat loss and possibly climate change. It likes wild flowers, red clover in particular, and lives now only in areas where traditional crofting practices are maintained. I want to see it, and invite my friend Matt to join me on a trip to the Outer Hebrides to find it. He agrees; he's up for anything is Matt.

From Oban the journey to South Uist takes six hours. We board the ferry in shorts and flip-flops, arrive in the dark with headaches and empty bellies. We're picked up by Sandy, South Uist's only taxi driver, a short, round man who speaks Scottish but gets by with English. Born and bred in Uist he hasn't strayed far from the Outer Hebrides and has never been to London. My name is written on a scrap of paper which he holds in his hand and points at – he can't pronounce the word 'Kate'.

Sandy drives us haphazardly down the middle of the road that rolls right out to the sea. We tell him we're here to see the great yellow bumblebee. He's never heard of it, but teaches us Gaelic for 'bee' as well as 'please' and 'thank you', and tells us to go down the machair. *Machair* is Gaelic for the type of wildflower habitat that clothes the coast of the Outer Hebrides. There's a particularly good stretch in South Uist, which is why we're here.

We travel for ages, eventually pulling up at the campsite on the far south of the island. We pitch our tents in the dark, drink beer and eat crisps before stumbling to bed.

In the morning we wake to white sand and giant sky, a landscape I had never seen or imagined before. I've never seen such beautiful beaches. They're perfect, untouched, littered only and occasionally by an abandoned stone cottage or a barnacled rowing boat. Everything else is seaweed and razor clams and slate and rock pools, plus the heady smell of sea. We fill our lungs and dip our toes. We feel like we're at the end of the earth. No one can get in touch with us as we walk, sometimes in silence and other times sharing stories and making each other laugh, me pausing to stare at patches of red clover or knapweed, towards the next island on the Hebridean archipelago.

We cross the bridge to Eriskay and happen upon a huge patch of machair. Matt goes off to meditate; he has downloaded an app on his phone and sits on a bench looking out to the Atlantic, following soothing instructions played out to him. Sometimes I look over and see him stretching or chanting. Other times the words of the Meditation Phone Man float over on the breeze.

You are important.

I walk around, looking at flowers. There are good numbers of the moss carder bee, *Bombus muscorum*, the strain that has a wonderful burnt-orange coat, rather than the one that looks similar to the brown-banded carder, *Bombus humilis*, which looks similar to the common carder, *Bombus pascuorum*. My interest in bumblebees wanes when I have to go looking at tergites and leg hairs and face length just to tell one closely related species from another. I'm happy to see a burnt-orange moss carder and know I have seen a new species without having to trouble myself with detail.

There are no great yellows.

Matt continues to meditate. I sit among clover and look out to sea. Oystercatchers call to each other endlessly, flies buzz, bumblebees hum. I try hard to bottle the moment.

Eventually Matt returns, feeling better for his meditation. We walk on as he sings Aerosmith's 'Love In An Elevator'.

Have you seen your bee?

Not yet.

What about that one? Is that it?

No.

How do you know?

Because it's a moss carder bee.

What's a muscarder bee?

Etc.

I try to describe the great yellow to Matt but it's hard as I've never seen one. I just know that I will know it when I see it. Something will click. Telling Matt it's a large bumblebee with an unusually long and mostly yellow abdomen doesn't really help. How do you know you haven't already seen it? He says. Because I do, I reply.

We come to a smattering of houses, Matt telling me why 'Love In An Elevator' is the perfect rock song. We find the pub, the Politician, named after the 1941 *Whisky Galore* shipwreck, where 260,000 bottles of whisky ended up on the beach. There's a scrap of machair in the garden. Matt orders beer and food while I scour flowers for bees. Nothing. Am I doing something wrong?

We eat burger and chips and sink two pints of beer. We continue walking, by now a little sluggish. We get lost in the hills, find sheep, boggy areas, we lose the machair. We head back and find the pub again. I feel guilty about dragging Matt around and so we sit, again, with more chips and more beer. Matt finds the juke box. The locals are watching football; Matt plays Queen three times to annoy them.

The five miles back to our campsite feel like ten as the sky rusts around us. We see a seal and some birds of prey that might be sea eagles but probably aren't and I curse myself for not doing more research. Outside the campsite we join others up a little hillock to watch the sun sink into the blue. We drink Tennents Super and cheer dolphins. It's been a nice day but for the most part I feel we have failed. We have one day left to find our bee.

The morning is Sunday and the campsite cafe is closed. We sit in the doorway looking out to sea, and fill our bellies with dry cereal bars and crisps. We don't really know where we're headed, except that yesterday we took a left and so today we should take a right. We head out, avoiding the beach in case we get lost, and stick to the road. On the verge is a dead hedgehog mown down recently, and next to it is a great yellow bumblebee feeding on a tired bit of knapweed.

Are you kidding me, is that our bee?

That's our bee.

I can see now, it's different, isn't it?

It sure is.

Are you going to keep the hedgehog?

I want to keep the hedgehog, but we are flying home and I'm not sure if it's legal to carry such things on a plane.

The bee hangs around a while. I photograph it with my phone, so I can text the image to my friend Ben, who confirms my sighting: my first great yellow bumblebee, a slightly sun-bleached male.

What now? It's still early. We head to the machair anyway. One great yellow bumblebee isn't enough for me.

We walk for miles, down endless roads, at the sides of which we pause for crisps and water, me looking for bees. We take a road to the beach. On the left is a pub and on the right is our machair. It's separated from the beach by a little wooden fence; there's a sign directing us in.

We stand at the crossroads and decide to reward ourselves with the pub after the machair. So we turn right, head into the flowers. Matt goes off to meditate.

It's windy here, salty and barren. But it's beautiful and there are plenty of flowers – huge patches of red clover. I walk around a bit, looking at patch after patch. Nothing.

I sit down next to the freshest patch of clover, which is sheltered slightly by a pebbly dune. It's cold and I think of the pub. Then, ever so briefly, a great yellow bumblebee pops up from nowhere, feeds on the clover and disappears again. I watch another do the same, too quick for me to identify it, but its shape is so distinctive I instantly know it. I'm ready for the next one, which I follow after it has fed, and it disappears into the grass. Ohhhh. Machair is grassland, albeit wind-flattened, hardened, pebbly grassland. And its huge hummocks of flattened grass are sheltering bees, which emerge from them to feed and then return, instantly, to shelter. So this is how they survive here. They hide in the long grass, emerging on still days and breaks in the wind.

I look up from my little patch of clover and scan the horizon. We were told last night that we have been blessed with exceptionally good weather. The sky is cloudless and the water calm. But can you imagine, they asked, what it's like here otherwise? They tell us that the road to Eriskay is closed in bad weather, and that various islands are cut off from the world until the storm passes. You might take the ferry to Oban and not be able to return for a week. The people, buildings and trees here are squat, battered from years of wind and rain. But they have worked out how to live here. So too has the great yellow bumblebee.

Voices, commotion, people. Next door there are builders, estate agents, all sorts going on. I can't get into the garden because of the noise. I can't be alone when there are men on scaffolding, Juice Radio drowning out the sound of my starlings.

I open the back door and every fear is realised. There's a boy, can't be more than eighteen, hacking away at the buddleia. Most of it is gone now, in its wake is a view of broken fence panels, the bluest house, a huge satellite dish. I ask him if he's removing it completely and he tells me he is, but he's annoyed because it's taking so long to dig out and his boss paid him for only half a day. He keeps calling it a lilac. This bloody lilac, he says, as he hacks and chops and digs and bags up. The house sparrows are nowhere to be seen or heard. I hold back tears as their home is destroyed.

Are you paving it? I ask, trying to sound as nonchalant as possible. We're getting rid of everything, he replies, the builders are coming next week, I think. We've got to get it ready for them but I can't get this lilac out and my boss won't answer the phone. I look at him, this child destroyer, with maybe £50 in his pocket for prepping some land to be drowned in cement. He doesn't know what he's doing, what he's done.

I go inside, close the door on him. I can hear him talking to his boss on the phone. She's coming round, now, with someone else. They'll bag it all up together. Plastic bags, no doubt, bags for landfill. Dear house sparrows, I love you but I couldn't stop them putting your home in the bin. I couldn't stop them bagging it up and chucking it out, to rot anaerobically with half-empty tins of beans and broken irons. Will you forgive me? Will you ever come back?

I can't be here, in this flat, this garden, this street. I need to get out, be gone with all this. I pack an apple, grab my jacket and binoculars, leave my phone. And I get into the car and I drive and I cry and I don't even know where I'm going but I end up on the Downs and then I turn off to my little place where I walk sometimes, where I feel happy and safe, where I can be alone. It was here I gathered haws and saw my first yellowhammer, where I watch swallows and tree bumblebees in summer and cower under trees knocked together by autumn winds. Ashurst. It's familiar and comforting and fresh and alive. Maybe I can be calm here.

I park at the pub and cross the road, climb over the stile. My neck and shoulders stiff and sore, my body shaking. Yet as I land and my boots sink into the mud beneath me, the call of a buzzard rallies above. It's OK it's OK it's OK.

The sky is a picture postcard, the odd cumulus cloud mirroring sheep on the Downs. A crow soars, a cuckoo calls, a woodpigeon clap-claps into a tree. A woman stands, an apple in her pocket, her world torn apart. Again.

I wish it wouldn't affect me so much, that it didn't matter so much. The loss of land and gardens I take so personally. Why? Why care so much about little brown birds that live in a toppled tree? Is it the tennis court, the buttoned-up childhood?

I tramp through the field, to the path bisecting a hedge and stream, with each step a little more free, a little more angst left behind me. I'm at the top of a hill and the land unfolds like a present unwrapped. I'd last been here in early May and everything has grown. Cow parsley towers above me, nettles brush my legs. It's like being small again in the wilderness of Dad's part of the garden. Now there's barely a path at all – the stream edge and hedge have nearly knitted together, like closed doors, opening just for me and closing in my wake.

Gardens I loved and lost. Gardens where I learned to love, where I cut my teeth in the natural world, where I first saw sparrows and blue tits and worms and caterpillars. Where I scattered seed, made compost. Where, when everything inside was falling apart, I could go, I could be. I could dig soil and bash cow parsley, I could lie under the oak tree staring through its branches at the changing sky. I could cope. And if I could just speak to people, if I could just say that the only way I have ever dealt with anything is via this wonderful world outside our back doors, and that I can't be the only one to acknowledge this, then maybe, maybe they would realise how big a deal this loss is. The cumulative loss. Not of one buddleia but the loss of front gardens to driveways one after the other, the loss of mature trees and parks and ponds because councils can no longer pay to look after them. The loss of a high-maintenance hedge for a build-and-leave fence. The loss of native wildflowers, of caterpillar food plants and nectar and pollen. Of birds and bats and hedgehogs and frogs. Every day a little bit more, another chip, another slice taken from our collective pie. Who killed Cock Robin? We did. Who killed the sparrows? We did. When will we realise that there is nowhere else for these creatures to go because we have paved over every last drop of our country. That, once it's gone, we will have only ourselves. And how boring would that be?

At the end of the path I turn right onto a country road, my boots echoing in the silence as they scuff the tarmac. They are all I can hear for a while, before I tune into birdsong: chiffchaff, song thrush, blackbird, woodpigeon, blackcap, chaffinch. In the trees, singing as they do every day, while we sit in offices or in traffic jams or our houses having existential crises, or paying some kid to cut down buddleia which he thinks is lilac and who has no idea, will never have any idea, what that shrub meant for twenty little brown birds he never knew.

Past huge country houses with their huge country gardens, past the church and its ancient cemetery, the homemade signs advertising chicken and duck eggs. The road is quiet and flanked on either side by masses of lush green nettles – I scan them for caterpillars. Nothing.

I walk on, scuffing tarmac, eating apple. A male blackbird practises his song for next spring. The sun is trying to come out but it isn't ready yet. The wind is picking up. The footpath takes me to a farm, where I turn in and walk through outbuildings where swallows nest, to get to another path that straddles the edge of a field and a copse of woodland. A magpie stands staring at me.

In Hove they are bagging up buddleia.

The wind comes again. At first a whisper and then a howl. Summer has stolen the horizon; I can't see for cow parsley. The trees creak as my boots sink further into mud. I push on along the wind-battered path, past brave ringlet butterflies dancing in the undergrowth. There are more trees, more wilderness. Sweet chestnuts and a hundred oaks, the ancient willow hiding its chiffchaff.

The path takes me to another field, with a good bit of hedgerow and field margin, where skipper, small tortoiseshell and meadow brown butterflies dance in the breeze. There's space here, a huge rolling sky. Swallows chitter-chatter as they swoop to catch insects. Two yellowhammers jostle on a telephone wire while Cetti's warblers boom in the hedgerow. I wish I could live here, instead.

I walk for ages, take a wrong turn, become hopelessly lost. I don't care. I turn onto a road, my boots scuff the tarmac once more. I've been out for hours, walking along, listening to birds, looking, in vain, for caterpillars. Eventually I come out on the main road and find my way back to the car, hunger the only thing ruling me now.

At home I let myself in and tentatively let myself out again into the garden. The little boy has packed up home for the evening, the scaffolding is empty of men. I lean over the wall and see bags of buddleia lined up against the house, ready to be taken through to the bins in the road. The buddleia isn't gone completely, maybe they'll tackle it again tomorrow. Everything else is chopped back, destroyed, ready for its dressing of cement to stop and suppress. It's quiet, and it takes me a while to realise it's because the sparrows have gone. Even the rooftop gulls are quiet.

I haul myself up onto the back wall and stand on it to look over the fence. I'm ten years old again, heartbroken, looking at my childhood wilderness destroyed to make a tennis court. There are paved gardens either side of me, at the end to the right is a bit of lawn and the smoke bush; to the left is decking and fake turf. I can't see anything else but there's no habitat now for house sparrows. I've failed. Everything I've worked for is pointless.

✿❀✿

When it snows, when it really snows, everything is silenced. Everything is stopped somehow, hidden from view, put on pause. A cat streaks silently across an unseen lawn, a bird searches, in vain, for food. But there's nothing else. There's no wind, no flutter of leaves. No cars, no buses. No birdsong, even. August, but winter again in the garden.

I've fallen out of love with it. This tiny bit of earth sandwiched between cement and roads, each building housing nine or ten people. Tenants moving in or evicted, landlords making a killing. Dogs barking, children screaming. Why did I do this? Why here? I could have moved down the road to a house with a bigger garden, a proper garden. I could have lived outside the city. Why did I chose something that was already broken and that would continue to be broken? Who was I kidding that I could fix this?

I sit in my corner drinking tea. I can hear a neighbour, two doors down, pottering in her garden. If I stand up she'll see me and I can't face that now – awkward introductions, garden chit-chat. Her next door is smoking on the steps; above her people make breakfast in the kitchen – the smell of toast mingles with cigarette smoke as I hear the scraping of buttered knife. This space isn't enough. It's not big or sunny or private enough. I hate being seen. I hate having to chat to the people two doors down, hear conversations between people five doors down. When I'm gardening I want to be alone. I don't want to be gawped at or waved at or talked to. Sometimes it's OK, but mostly just leave me alone.

It is, of course, a miserable grey day. Not even the sun could shine for me. The sparrows have been gone for a week and the void is huge. They'd only just started coming in. This was theirs only for a few weeks. Five weeks? But before that they were next door, in the holly to the left and the buddleia to the right, and now they are nowhere. The silence is everything.

Trellis. I need trellis. If everyone else is cutting and destroying then I need to create, to block them out. I need to walk out here without interrupting people smoking on their steps, without being talked to, without witnessing the destruction of a habitat. Trellis would give height to the garden, make space for the roses and clematis to grow into, keep neighbours out but bring more wildlife in. Maybe, in time, the house sparrows will come back. I price it up on my phone. I can have it for less than £200 if I do it myself. Can I do it myself? I've not put trellis up before but it can't be that hard, can it? If I can take decking down . . .

I get up, buoyed slightly at the prospect of privacy. Or some of it. I weed a bit, bindweed and avens, leaving ivy-leaved toadflax in the walls, willowherb in a small clump. I tie in and prune and plant. I move foxgloves. I divide and replant lamb's ears and common bistort. I figure if I grow more plants around the greenhouse then it might look nicer, less out of place, less huge and awkward. But, really, it has never looked good. It needs to go.

I stand and survey my land, defeated. The back border looks terrible, suddenly. The plants have grown together too much and there are big muddy gaps where I'd sown seeds that haven't done anything. And I didn't stake my poppies, which are white but should be red. The pond hasn't filled up, despite two more days of rain. Willowherb has popped up in all the wrong places and I can't move it because I want elephant hawkmoths to lay eggs. And they haven't. Slugs and snails have been on the rampage. The last of the sunflowers I planted stands ravaged and alone. I feel like a failure. I want to dig up the whole thing and start again but I can't.

I tear the pathetic excuse for a sunflower out of the ground. Its stem is thick, strong and spiky, it would probably have survived. But who wants one sunflower in the middle of a border? I planted nine. It's better to just not grow these things.

I glance at the pond. Two pond snails writhe together at the water's surface. I sit for a minute and watch them. The water reflects the sky and I see starlings and herring gulls on the rooftops. A calmness washes over me. At the pond edge is a bit of valerian I'd pulled out of a wall somewhere and have been rooting in the water. I lean over and pull it out to see if it's ready for planting. Yes, but there's a clear sausage of pond snail eggs attached. So many steps forward and so many back. Then forward again. Will it always be like this? What will be destroyed next? I pray for years of rain and sunshine so this garden can *grow*. Sustain itself, sustain its wildlife. There's no buddleia but there are baby pond snails. There are no sparrows but there are baby pond snails. Baby steps. Baby snails. Baby love and baby hope. The two things that keep us all going on this crazy world that, for some incomprehensible reason, keeps bloody turning.

✿❀✿

My garden was grey but now it's green. When the sky is blue I can convince myself I've done something important. I can sit on the lawn looking at bees or top up the bird feeders or watch dragonflies lay eggs in my pond. It's easy when the sky is blue, when the garden is green. But it's not enough. It's not enough for me and it's not enough for anything. I can rip up decking, grow plants, put up nest boxes. But there's nothing I can do about next door. I can never stop someone destroying a habitat, taking down a home, chopping something up or paving over it. This garden I've created might be insurance against further loss but it won't reverse it. It can't. And look at it: a little green piece in a jigsaw of grey. The sparrows won't survive here alone.

I run along Brighton's streets looking between gaps in houses and over roofs at what lies beyond them. Sometimes I catch glimpses of trees and life; mostly I see sky – evidence of more gardens lost beneath tarmac. Space enough between large houses to build an access road to a car park or a mews; homeowners and developers made for life; gardens lost for ever.

A school extended. Shrubs and trees cut down, walls put up. A church, which long since lost its congregation, converted into luxury flats. Mature trees cut down and replaced with low-maintenance shrubs with little or no use for insects, seemingly green but not even close. Plastic trees, plastic grass or the next worst thing. A razed front garden; its contents, still lush and green, lie in a skip on the road, roots angrily facing the sky. A child sits on steps before a paved front garden playing with a toy car.

Piece by piece, our cities are being turned into luxury flats and car parks. Developers take the money and walk away, the people none the wiser. Councils are allocated less and less money to maintain parks and green space. Will they be sold off too? Where will the children play? And when these children are in government, what will they know and what will they save when budgets are further cut? Luxury flats? Car parks?

I sit alone in my flat, the garden, still silent, beyond the back door. Fuck it. I dig out my bag of haws, which has been in the freezer since Mum visited last autumn. I had planned to make 'hawsin sauce', a sort of ketchup that goes nicely with chips. But I never got around to it. I empty the bag into a washing-up bowl of water. In a few days I'll sieve the contents, separating floating flesh from heavy, hard, brown seed. I'll put the seeds into the pockets of my running jacket and gradually, over the course of the next few weeks while running at dusk in autumn rains, I'll push those seeds into Brighton and Hove's forgotten soils. I'll choose retirement and nursing homes, hospitals, schools and communal flats; places where landscapers trim back and mow for tidiness but rarely bother to dig out errant seedlings. It could take two years for my haws to germinate – twenty for anyone to notice what I've done, but their legacy could last a hundred. How many moths will lay eggs on a hundred hawthorns in a hundred years? How many birds will gather their caterpillars for their young, the haws in autumn? How many developers will rip the plants out and pave over them, the people none the wiser, tramping in their wake? Who knows who knows who knows.

Autumn

The boy who cut the buddleia and razed next door's garden was mistaken; it would not be paved after all. The removal of all plant material revealed that it was mostly paving stones anyway; just three deep, empty beds remain. I jump the wall and throw wildflower seed onto the beds: teasel and willowherb plus some cornfield annuals for bees. Try to stop me.

I can't stay, not here. This isn't working. I look at property online, scour websites and maps and find the biggest gardens in Brighton and Hove, not too far from the sea. The largest are on the train tracks. I take trains to Portslade and Preston Park, glimpse large gardens through the window. Old apple trees, climbing roses, long-forgotten swings. Sometimes I view a house. Mostly the gardens are disappointing: paved, neglected, deathly quiet. But I find my dream home on Wilbury Crescent. It has a huge south-facing garden with two apple trees, a pear and a cherry, a long lawn and a crumbling shed, and masses and masses of potential. All mine for a million quid.

I take the greenhouse down; it was never right. In its wake is bare soil, brown against the lush green of the rest of the garden. I clear away tomato haulms and plastic pots, bottles of plant food and bamboo canes. I dig the soil. Now what? I call Helen. The remains of the turf she laid in summer is still piled up in her driveway: Do you want it? It's dead. Yeah, I want it. I load rolls of mud into the car, take it home, through the flat. It's brown and thin. I rake the earth where the greenhouse stood, make everything level, lay this last bit of lawn. It will grow, everything always grows. Early September, it will be green by May.

I buy trellis and paint it on misty afternoons, starlings shimmering above me like schools of mackerel in the sea. The paint is a sort of blue-green. On the tin it's called Gentle Sage, other brands call it Willow. It's the colour of globe artichoke and lamb's ear leaves. I put the trellis up on my own, a bit wonky but serviceable. I plant Clematis 'Bill Mackenzie' and watch it weave into the space.

I spread compost on the soil, release plants from pots, move things around, divide and replant, take semi-ripe (nearly ripe) cuttings, bury the first of the autumn bulbs. Why am I doing this if I don't want to stay? Habit, I suppose. For the house sparrows, I suppose. Feed the soil and everything will follow, I suppose. Feed the earth, the detritivores, the centipedes and beetles, the roots of plants that will flower and seed and fill trellis and protect birds.

The garden is transformed again, better and bigger than before: a small, irregular-shaped, half-dead lawn surrounded by borders and a beautiful little pond. Climbing roses, clematis and honeysuckle slowly colonising the fences and trellis behind. Scraps of this and that, useless now but that will grow into the space. Elephant ears, white deadnettle, cranesbills and foxgloves, plants gathered from tree pits and cracks in the pavement and road, rooted in water and planted out. Serbian bellflower, which seeds all over the city, now grows with red valerian seized from a wall. Linaria taken from the Sackville Road, honesty from a supermarket car park. It will be spring again. It will be spring again.

✿❀✿

Granny's wedding dress hangs in the living room. Like a ghost, like Miss Haversham. What's it doing there, I ask Mum, why do you still have it? Posterity, she says, and because I want my mum with me on my wedding day.

It's been there for two weeks. Every time Mum walks past it she touches it, touches its folds, its traces of skin and hair, of sweat, of Granny. It smells musty and has faded from white to a sort of cream-brown. Mum digs out a grainy black-and-white photo of Granny and Grandad's big day, the two of them standing side by side with matching sets of parents beside them. The photo reveals the dress has since been altered. In those days, says Mum, women often adjusted their wedding dresses and wore them for other occasions. Granny didn't wear this dress once but many times.

It's nice to be away from home, with Mum. She lives in a part of Solihull I don't know well; she moved here only two years ago. Beneath Birmingham airport flight path so it's noisy and full of planes but there's space here: countryside, fields, hedgerows, nettle beds, life. Wasps and house martins nest in the roof of her house and she gets nuthatches and goldcrests on her feeders. At the end of her garden is a huge purple beech where mistle thrushes, woodpigeons, tits, all sorts hang out among the branches. I watch them for hours. I want to climb the beech but Mum says that, at thirty-five, I'm too old. At the end of the lane is a large fishing lake she walks around every morning. When I visit I run, lap her three times and run back. The lake is managed for fishermen but there's plenty else for me: kestrels, great crested grebes, a black swan.

Wedding bells ring through the village. I'm on flower duty with Trudi, who came into my life recently and unexpectedly, and who's now being thrown in at the deep end – a huge family wedding. Every table in the village hall will be decked with home-grown blooms, all lovingly assembled by us. On top of that I have the small task of walking Mum down the aisle and making the father of the bride's speech. I'm terrified but I can't let on. No one must know.

I get up early to run around the lake, the pebbled path crunching beneath me. I feel calm here, safe ahead of the Biggest Thing I've Ever Done. The trees are holding on to their leaves still but there's been a shift, a shrinking back, a collective breath held, a belt tightened. That which I love is disappearing for winter. I try not to think about it. I try to think of the flowers, the celebration, the happiness of Mum and her new husband, of seeing her brother, my godfather. I stop to watch herons fishing for breakfast, the first of the autumn's redwings gathered in the field beyond the hedge. Sloes are nearly ready for gin.

I always run clockwise around the lake; Mum walks anti-clockwise. For no reason other than that's the way we go. I'm on my third lap when I find her, a smiling thing all caught up in her own emotion, her own trepidation about the enormity of the day. Mummy! She hugs me and tells me I'm the first to see her today; I feel proud and privileged. I've never seen her so full of joy. We chat about the herons, the flowers, her soon-to-be-husband, Pete, the order of the day. We part again as I run and she walks, lapping the lake in our own way.

Later Trudi and I arrange flowers: *Verbena bonariensis*, dahlias, fennel, erysimum. I spend three hours in the hairdressers, come back, put a dress on, walk Mum down the aisle, make a speech, eat, drink, dance, celebrate. The day ends drunkenly, as these things do, and in the morning I wake, fuzzyheaded, in the wrong bed, to church bells and nuthatches, flowers strewn about the village like confetti. I wish, more than anything, that I didn't have to go home.

✿❀✿

Most mornings I stand on the kitchen doorstep drinking tea. I look at the sky for interesting birds, keep an ear out for house sparrows that might yet return; other times I venture out and am lost for hours.

Today the garden takes me. It's cold suddenly and everything is frosted. I crouch down before the far, newest, border, where the greenhouse stood, and conduct a sort of audit. White deadnettle has germinated from seed I took from a car park earlier in the year; its fresh green leaves poke out of the ground. The honeysuckle cutting from Mum's is still tiny but has buds ready to unfurl when temperatures increase again. The bracken is dying down for winter. There's green alkanet, which has self-seeded in from next door and which I'm too polite to weed out; cuttings of box which I've planted in the soil to grow on for a bit before I decide what to do with them; elephant ears ready to expand into the new space – only one of its paddle-like leaves has turned red for autumn. The perennial wallflower and foxgloves are doing marvellously – we will have a huge display next year; there are bits, scraps, of hardy geranium at the front, the remains of a euphorbia, a half-dead clematis, some kale I'd forgotten about, bits and pieces of this and that. Between the plants are masses of love-in-a-mist seedlings which I gathered from this summer's plants and scattered everywhere I could. It should look nice when the time comes.

At the back *Echium pinnata* has filled a corner. The apple is losing its leaves. The cosmos has finally flowered and the linaria, which I took from a pavement crack, is thriving. The valerian is doing well; the crocosmia, which Helen promised me wasn't orange, is orange. There are gaps where I've planted alliums, the pond is filling up.

I feel a bit useless, out of sorts. I return indoors, re-boil the kettle and make more tea, return to my step where I'm warmer and where I can poke my head out of the door to see what arrives when my back's turned. The tea steams into the sky. I stand on the step and sip.

House sparrows. I hear them before I see them. In the holly at the end to the left, I think, now the buddleia is gone. It's been ages but there's a big hungry gang of them. One flies to the feeder just in front of me and does an about-turn with a *thuuuuur* as it makes its escape. Oh my heart, they're back, they're back. They're cautious again now but that's OK. As long as they're here, as long as they're willing to give the garden a second chance. Their cheeping fades in the distance as they go off again.

It doesn't matter, I won't stand here all day, they'll try again later. I fill up the feeder in anticipation.

There are more birds, other birds, the cold will have brought them in. I hear a robin pink-pinking at something, and a blackbird, which is new. The blackbird is pink-pinking as well. At the robin? Yes: the robin then flies over the garden, quickly pursued by the blackbird – a territorial fight.

The blackbird moves off and the robin flies to the smoke bush at the end to the right, disturbs a gang of tits. I drink tea on the step, stretch my runner's calves. Wait. The robin flies in, lands on the sunflower feeder at the back, takes a seed and flies off again. And then another bird, a blue tit, I think, and then a goldfinch! All new. All in a dash. But it's promising. The apple and clematis and roses and honeysuckle will provide them with shelter in time.

I boil the kettle again, make tea again. Empty the dishwasher, wipe the worktop. The back door remains open and I listen for song, the chatter of a goldfinch, the burp of a blue tit, the watery, staccato trickle of the robin. Instead I hear the blackbird's gurgly song from beyond the wall. And then it stops gurgling and hops onto the wall, and I realise it was only next door. It sails down on to the ground feeder, a young gun perhaps, establishing a new territory. I strain to see it but can't and it emerges a few minutes later from the pond. My very own blackbird, hey. Finally, the garden is a garden.

The sunflower feeder is rarely visited and I'm pleased to see it in use at last. I draw a line where the food is with a marker pen, check the seeds haven't become mouldy. If there are gangs of tits and charms of goldfinches and an angry little robin then the seed might yet be eaten.

I fetch my drill and take down my bee hotels, full of cocoons now, rather than grubs; little packets of box-fresh bees, which will sit winter out in safety and emerge when the apple blossoms. I don a coat and hat to ward off the cold and sit cross-legged on my living-room floor with the back door open so I can see sparrows on the feeder. I tease cocoons from the hollow plant stems. Spiders and woodlice spill out everywhere, there are a couple of snails and some strange things I can't identify. There are more cocoons in the fancy wooden hotel and I manage to tease a row out using a flat-headed screwdriver, but I'm scared to touch the rest of them, scared I'll damage the bees with a clumsy slip. I leave them in there, gently brushing frass from around them with a paint brush. I count sixty-three bee cocoons, an improvement on last year's forty-seven, and with the happy addition of leafcutters to complement the red and blue mason bees. I transfer the hollow stem cocoons into the release chamber in the fancy wooden hotel, sweep up errant spiders and woodlice that are trying to find new homes in my living room. I throw out the hollow plant stems now, these have definitely had their day. I carry my wooden hotel with its precious cargo to my shed, where it will remain cool and dry for the rest of autumn and all of winter. A box of bees and sunshine not to be unwrapped until spring.

Winter

It's 1985 and I am four. Mum and Dad's friend Charles has given us a bird box. It's for tits. He told us to put it up in a north-easterly direction. He chose a bit of fence by the back patio doors. The bit of fence above the hydrangea, where a scrappy bit of cotoneaster grows. Mum says the birds won't come so close to the house but Charles says they will. We'll see! says Mum.

The box remains empty for so long that Mum and Dad forget all about it but suddenly, eventually, it's being used. It has attracted tits, just like Charles said. The tits are blue tits. My sister Ellie is too young to really appreciate them but Dad makes me stand still and wait for the birds to enter and exit the nest. I am better at being silent now I'm four, better at understanding why and how to be still. But it's hard, waiting in the shadows behind the patio window, craning my neck to catch a glimpse of them. I see the parent birds journeying to and from the nest. They look the same but we think the male has more of a Mohican hairdo. They look frazzled and tired, unkempt, not looked after – Mum makes a joke about parenting. The birds bring big green caterpillars from the trees by the garage, take faecal deposits from the nest to distribute around the garden. Sometimes a tit will perch on the top of the nest box, sometimes we leave crumbs and bacon rind for them to eat. I like watching them. They go in and out, in and out. As time goes on they become more used to us and if I stand beneath the box I can hear the chicks cheep-cheeping inside. If I make a noise they cheep louder; they think I am a bringer of caterpillars.

It's a nice sunny day and the patio doors are open. I'm on the patio, wheeling around on my toy truck. The tits don't seem bothered by me. Dad is skulking in the corner. What are you doing, Dad? I'm watching the birds, he says. He's watching them intently, learning the different routes the male and female take, timing the gaps between food deliveries and faecal drop-offs. He has worked out that they gather most of the caterpillars from the border between the gravel path and the driveway, that it takes up to five minutes to find enough food to bring back. He watches the female come with a beak full of wriggling green grubs, watches her disappear into the box, feeding her hungry babes on the other side of the hole. He notes the male, waiting in the sidelines, he, too with a beak full of baby food. The female pokes her head out and takes off. Dad notes the tree she flies into as the male sails down to deposit his parcel of food. Dad's poised, waiting, me none the wiser on my little red truck. The male flies off, into the same tree as the female.

Quick! Large hands scoop me up and I'm transported in the air to the tit box. He puts me on his shoulders and opens the hinged roof. Five yellow mouths open to greet us. Five naked pink bodies and barely open eyes. Can you see them? Can you see the birds? I'm speechless, I don't know what to say. Dad closes the lid as quickly as he opened it, takes me off his shoulders and places me down again. Did you see the birds? Did you like them? I look at him and then back at the box, now knowing what's inside it. I am troubled, almost, with this new information, that baby birds do not look like birds, but tiny monsters with giant, gaping mouths. The female flies back with a mouthful of caterpillars and we watch, together, as she feeds her young, unaware of the recent intrusion. She flies off again and the male returns, he, too, unaware. I want to see them again but Dad says no, we have to let them raise their young in peace. That was a one-off, just to see what was inside. I spend the rest of the day thinking about what they're doing, what they're eating, if they're sleeping, if they've grown any feathers. Five little birds in a box outside the back door.

When I erected the trellis I had to move the bird feeder. Not the one the sparrows visit, which I refill every other day, but the one on the back fence, the one full of sunflower seeds that the sparrows ignore, the one with the food that goes mouldy and has to be replaced every three months because it's never eaten, even though I only ever half-fill it. Over the last few weeks I've watched blue tits, goldfinches and the robin fly in from the smoke bush in the garden at the end to the feeder. One by one, each takes a sunflower seed and dashes off again to shelter. I had hoped this might be the beginning of something – robins are fairly antisocial beings but goldfinches are nearly always seen in big groups, or charms, and blue tits hang out in winter with other tits in 'roving groups', which basically means they eat together. I imagined a swinging feeder with goldfinches and tits fighting over food, imagined the pecking order, the fledglings being taught where to feed. It wasn't to be. Shyly, the birds have stayed in the smoke bush and only popped in to grab a seed and fly off again.

But I open the back door now and the bird feeder is swinging. There's a commotion in the smoke bush. It's tits. Lots of them. A roving group all burping and fluttering and hanging off the branches which bend under their weight. Tits. Then there's a flutter of wings and I catch sight of the robin leaving the garden, and I realise I opened the door on a party of birds, although probably, if the robin was involved, a party involving a fight. I walk to the feeder and my tide-mark is a whole inch above the new level of sunflower seeds. And I am gloriously, fanatically happy because I know, just as soon as those leaves and stems inch their way into the trellis, that the birds will no longer need the smoke bush beyond the back wall of the garden. They'll have shelter enough here and they will visit more often. In ones and twos and sixes and sevens and huge roving groups or charms or whatever. The birds want to be in my garden, they just need to be able to hide while they're here. And they will, they will.

The next morning I get up early and drink tea in my coat at the kitchen door so I can see the tits and goldfinches that got through a feeder of sunflower seeds in two days. I'm too early. It's still dark and only the robin is around. I can hear her warbling away in the twilight. She's on the wall, pottering about beneath the trellis, singing away as she ferrets for food. She's a violin concerto, a trickling fountain, a dribbling mountain spring. And those little pips, the tinny echo of someone practising their scales on a glockenspiel. I wonder if she's actually doing scales, practising. She's probably a he, gearing up for the big showdown, the battle for territory and females, fought and defended by the quality of one's song. She hops around a bit, jumps on the feeder, takes a sunflower seed, returns to her spot betwixt trellis and wall. She's well hidden there, I can see why she likes it. Just you wait until the leaves come, girl.

The sparrows visit the feeder closest to the door, happily, but if they catch a glimpse of me, even behind the window, they fly off. It would be just like them to come at the same time as the tits and then sound the alarm, emptying the garden. I'm well prepared: I stand in darkness, dressed in dark pyjamas and a dark coat with my hood up. I am a hide.

The blueing sky is full of gulls returning from the seafront. Then: whoosh! A flash of rustling of leaves. I look up at starlings flying overhead, returning from the pier. They're bang on time, thirty minutes before sunrise, back to their street, to their rooftop. Two gangs, three gangs, four. The robin and herring gulls fill the gaps between starling flypasts until all goes quiet. The other birds don't come and I give up.

But later, as the sparrows descend on the feeder nearest the back door, I try again. I inch forwards and peer through the window. The sparrows fly off, of course, alarm calls sound, but I can see the feeder beyond and it has something on it, something big. I tease open the back door. One big thing flies off and the other turns to look at me. Collared doves.

I stand at the door and watch; the one on the feeder returns to feeding and its mate comes back to the garden. It comes in stages: first the roof and then the trellis.

Finally back to the garden. They perch, the two of them, balancing on the tray designed to catch seeds from the feeder, one beak each in a 'port'. The weight of them; they'll have the trellis down.

Dressed for work in a smart suit of pink-brown with a white-edged black collar and matching white-ringed black eyes, they're smaller than the woodpigeon and have a faraway stare. I've seen them before but not on the feeder. They come in, usually with the sparrows, sharks to their remora fish. They perch on chimney tops and *coo-coo-coo* all day long, an alto to the starlings' nonsense soprano.

They arrived here from the Middle East in the 1950s, first breeding in Norfolk in 1955. Their spread has been astonishing – they reached Scotland by 1957 and had crossed the Irish sea by 1960. Their population exploded until around 1996. It's no wonder: they have up to five broods every year. Sometimes they start a new nest before the young from the previous one have fledged. Two eggs at a time. They typically nest until October but in mild weather they can keep going and have been known to set up camp in municipal Christmas trees, preventing councils from ending festivities. I don't know where they nest around here; they could be nesting now. They lay eggs on the flimsiest platform of twigs; it's a wonder they manage to raise young at all. But they do and here they are, swinging from my hanging sunflower-seed feeder, bobbing their heads as if for apples. I watch them for a while, then head out and see them fly off. Their wings sound rusty and in need of oiling. They land on a chimney and watch me unscrew the seed tray from the feeder, retrieve the bag of sunflower seeds and fill both hanging and ground feeders, so there's no need for them to perch and swing for food. They leave their chimney and fly off who knows where. They'll be back in five minutes. Back to the ground feeder and not the hanging seed tray. Back to the garden and not the trellis. Collared doves.

January. The compost bin is full. I blink and the bird feeders are empty. Some days it's mild and the buds threaten to burst, other days it's freezing. Blowing hot and cold, it's terrible for wildlife. It doesn't know where it is, what it wants to be doing. Nest or rest? The not-knowing wastes energy. I worry about the insects – already I've seen a queen wasp and a small tortoiseshell butterfly – hedgehogs out looking for food, birds carrying nest material. It's a good twelve weeks before the sap rises.

There's never much to do in the garden in winter. There's no weeding to do, no tying in, no deadheading. I could chop everything back so it's all ready for spring but then there would be no hiding places for insects, no wild seeds for passing goldfinches. So I leave everything as it is, to protect and cosset the wildlife. I'm itching to clear the decks but I can't, not yet.

There are other things to do. Structural things. Planning things. I pull down the back seats of my car, line everything with cardboard. I drive to stables near the racecourse, where there's an enormous pile of horse muck. I dig deep into the steaming centre for my crumbly, sweet-smelling reward – the best bit, which is full of nutrients and soil conditioners and other traces of Wonderful Things that will make my garden grow better. I fill my car boot with old compost bags full of muck, strap more into the front seat and squeeze two into the footwell. I drive back as the sky burns over Brighton, my bagged passengers of horse shit steaming gently around me. I hope I'm not stopped / I hope my car will recover / I hope I don't see anyone I know. I drive home and relieve the car of its bags, carry them through the flat and empty them onto the borders. Everything here is hidden now, tucked under a little blanket of crumbly, sweet-smelling manure. Brandling worms and earwigs tumble out with it; I watch them scurry quickly to shelter. I tuck muck around the honeysuckle, which is still only 10cm high but which has everything it needs now to *grow* in the coming year; the apple tree, the clematis, the roses. I brush it off the tips of daffodil leaves which have just pushed through the soil, off seedlings of honeywort and love-in-a-mist. Over the next few weeks the worms will drag the manure into the soil, mix it together, recycle it into worm casts. The new blackbird will pick through it looking for grubs. Already, it's a habitat. And then when the sap rises . . . it will be like fireworks going off, both in the garden and my heart.

✿❀✿

I crouch in my corner in the rain, listening to droplets on leaves. I like it here, no one can see me. My jacket pulled over my head I blend into the surroundings. I am rained on like a leaf.

Rain on leaves is a summer sound, the water pattering in blobs, the leaves bouncing on impact, the drops splashing, disintegrating into something else, forming another drop or a bigger drop or a puddle, a pond. It soaks into earth and is taken by roots and stems, journeys through a plant and into leaves, evaporates back into sky, collects in clouds, falls again on leaves.

Does rain make less noise in winter? Fewer leaves to land on but less to break its fall – does it puddle more? Splash more as it lands? Today it's heavy and it bounces off my raincoat, adding to the pat-patter and the bedraggled but nourished state of everything, including me. The leaves and I are in a halfway house between winter and spring. We make tentative steps, unsure, not yet willing. We dip a toe, we squeeze but don't pop.

The leaves are children, perfect and precious. Uneaten, unmangled. The lungwort is the greenest it will ever be, its disease-esque markings a gorgeous shade of cream, its hairs erect, on guard. Primroses in tight clusters, their ribs and veins holding onto rain. Bright green hellebore leaves poke through the soil making the darker ones look old and tired. Make way for the new, they say.

The honeysuckle's purple-green leaves catch droplets in their folds. My little cutting, that I took from Mum's in late March and carried to Brighton in a bag of water like a fish. I cosseted it and then forgot about it as other priorities took over. Then I moved it around too much. But when the greenhouse came down and the trellis went up I found it its Forever Home – the spot beneath the trellis in the shady bit of the west wall. It would grow up its little ladder and then north along the framework to the sunshine. Honeysuckle likes its roots in shade, likes a lot of moisture, but its leaves like the sun. It will be happy here. It's a few centimetres off the ladder but it should reach that in no time. Will it reach the trellis by the end of the season? Move along it, find the clematis coming from the other side? Does it have sufficient roots? Sitting now in its bath of manure, it's ready and raring to go. Just another few weeks and it will head skyward, wrapping itself around the trellis ladder, a worm in the soil, a spiral staircase, feeling its way, clockwise, to sunshine, to what it can hold onto. One stem of 'Frances E. Lester' is tied into trellis, the other is less than a metre from it. But it's in leaf, sort of. Little leaves at the end of a stick stem, buds unfurling, winding.

Sedum spectabile pokes through the earth in tight knots like unfurling Brussels sprouts. The freshest grey-green, each leaf tooth snagging a droplet; leaves that look like flowers. What else pokes through? Lychnis, oriental poppy, crocus, cranesbill. Good to see you again. Felty leaves of phlomis gather, foxgloves bustle, alliums stand tall. They're Victorians ready to promenade. There's my Chelsea bearded iris and next to it should be the verbascum that tried to die – beautiful thing, did it succeed? I hope not. I shuffle on bended knee to where I think it might be, brush away manure. There are leaves, little and soft. But not really in a rosette like a verbascum. I think they're lamb's ears.

There's so much more that's yet to be seen. So much beneath the soil, so much I haven't noticed. And every day now there's a new thing, a leaf for rain to christen, a bud, a stem. Loveliest of all, in this rain-soaked garden in its first proper spring, is the forget-me-not that's seeded in from next door. Not a little but a lot. Tight clumps of veined rosettes, long leaves spreading out, taking up room, feet splayed. There wasn't a single forget-me-not when I took the decking down and now it's making itself known, muscling in on the space.

I break off artichoke leaves which have fallen onto new growth. Out with the old. Make way for the new.

Spring

My ugly duckling is shaking off her grey down, the decking and stones, the mud, the thirty-year winter. Now when the wind blows it streaks through long grass, leaves of 'Frances E. Lester' flutter in the breeze. There are buds and stems weaving into trellis. Adult feathers poking through.

I sleep with the window open so I can be woken by the dawn chorus, bedding down each night as foxes still screech into the darkness. I pretend I'm eight years old again, in the spare room at Granny's house. I see her at the doorway telling me to leave the window open because the birds sound so wonderful in the morning. OK, Granny, I say, OK. And I go to sleep not really knowing what she's going on about and I wake up and there's an orchestra playing in my heart.

I wish I could remember more. Did she come in and listen to the birds with me? Make me tea? Sit on the end of the bed?

It's cold but the duvet warms me. Usually I sleep soundly, usually I miss dawn and wake only when the neighbours rise and tramp above my head. Today I'm woken by rain, the pattering on this piece of guttering or that, the repetitive ding-ding of water falling from somewhere onto something but I can never find out what. I'm disappointed by the rain because I want to plant peas. But also I'm disappointed because the dawn chorus is rubbish. I listen hard between the raindrops and can hear my house sparrows chirping, the blackbird a few doors down. No robin, she's long gone, and no tits or finches either. It's too early for the starlings, they're still on the pier. It's just me and the house sparrows and the blackbird and the rain. I drink tea, deflated, the ghost of Granny telling me to keep trying.

And then I hear it, a twittering, a beautiful sound like a tingling bell. And I recognise it but can't place it. Chaffinch? No. Dunnock? It sounds so close but it would never be in the garden, it couldn't be in the garden. And I'm so sure of this that I forget about it and get up and potter about and make more tea and eat porridge and then I open the back door to take the kitchen scraps out to the compost heap and there's a dunnock on the steps, hoovering up spilled seed from the feeder above her. And I freeze and stare at this dunnock, who carries on about her business and completely ignores me, as I did her while she was singing. And she hops on the steps, into the box bush, eating a bit of seed here, investigating there. I step closer and she flies into the smoke bush at the end and launches into her incredible flutey song while I stand in my pyjamas in the rain, holding my kitchen scraps, laughing. Laughing because there's a dunnock in the garden. Laughing because I'd disregarded her just as she disregarded me, laughing because in the last two weeks I've seen a bird on the camera traps that was always out of focus or in the shade and I know it was this dunnock. My walled, immature garden is about five years from providing the shelter that these so-called shy, hedge-loving birds apparently need. And either side of me is so much paving. But I know now that she's been here for a couple of weeks and has made herself at home. Will she nest here? No, the roses are sticks, the honeysuckle is two inches tall, half of the lawn is still mud. Will she come in regularly and eat spilled seed? Yes. Will she bathe, drink water from the pond? I hope so.

The dunnock is small and brown with a grey underbelly and bib, and grey bits around its head. It looks like a sparrow but has a razor-sharp, slate-grey beak for teasing insects out of tight spaces. In some parts it's called hedge sparrow because it loves hedges so much. My dad has dunnocks in his garden because it's bounded by hedges. They hide in them all day and dart out to the ground feeders or hoover up spilled seed from the hanging feeders or they bathe in one of the baths dotted in the border. But the whole time they're out of the hedge they're looking around, nervously. They're happiest in the hedge, happiest where they can't be seen. You can look out to Dad's garden and the whole area beneath the hedge can be littered with dunnocks and you don't see them because they're grey and brown and they blend into soil and shadows. A blue or great tit in my garden? Yes. A robin or a blackbird or a collared dove or a woodpigeon? Of course. But a dunnock? I'm surprised. This dunnock can obviously see the shape of things to come, can see the shelter developing. Has vision. Or maybe she's just really hungry.

I fill the hanging feeders for the house sparrows, who will knock seed to the ground for the dunnock. As I'm doing it a collared dove lands on the trellis, hoping for its own fill of spills. I scatter more seed in the ground feeder, seeing as the doves and the pigeons and the blackbird and now this dunnock are all having to share.

Before I go in again I check on my plants. Everything is the same except 'Jan Fopma', the clematis, has poked two stems through the soil. I planted her with 'Frances E. Lester'. Frances and Jan are raring to go and they will look beautiful together, they will provide shelter together, they may one day house my dunnock together.

My former home in London has been rented out for two years, and the garden is now an overgrown woodland. The dog rose is in the sky, the summer jasmine on the bicycle path beyond the back fence. I clip everything back, chop it all up and use the clippings to mulch the 'woodland floor'. It's a world away from the paved desert Jules and I bought seven years ago. I dig up self-sown Japanese anemones and dog violets to take back to Brighton.

When I lived here I worked on the communal garden, too. It was hardly a garden, just a closely clipped lawn, a couple of palm trees and a few raised beds with variegated ivy growing up huge trellised walls. We had a small 'makeover' budget once, and I was instrumental in getting native wildflower green roofs installed on the car-park vents. I planted trees: an ornamental cherry – not great for wildlife but free, from a neighbour, and the woodpigeons liked the fruit – and some silver birches. I sowed pollinator-friendly ox-eye daisy and red clover, planted *Verbena bonariensis* and sedums. I took cuttings of my climbers and squeezed them in, tried to make the place greener, more of a habitat rather than an empty space, green only in colour. I hung bird feeders and erected nest boxes, put in a compost bin. It was an uphill struggle. I argued constantly with the gardeners and other residents about what should be planted and how the garden should be managed. We like the bird feeders, they said, but we don't like the dandelions. We like the palm trees because they remind us of being on holiday (but do you need seven?). We can't give over the whole place to wildlife, that would be terrible, that would be messy. Why don't we plant pretty bedding flowers in those ugly green roofs? They wondered why the birds never came. It was like banging my head against a brick wall.

The flats and garden were managed by a management company, to whom we would pay an extortionate service charge each month. Some of this service charge paid for the maintenance of the garden, which consisted annually of: applying a weedkiller and lawn feed in spring, mowing the lawn once a week, bagging up the waste and throwing it into landfill. Occasionally there would be a budget for 'bedding': gaudy purple primroses in spring, petunias and pelargoniums in summer, pansies in winter. Nothing to see here, said the garden to the bees.

Each week the gardeners would come and I would watch them carve the place up with their lawn mowers and leaf blowers, bag everything into plastic sacks and take them to the bins. I spent months trying to convince them to compost the waste, not use plastic bags. Eventually, sometimes, they would leave the bags out for me, which I would empty myself into the compost bin, mix with brown material to stop it breaking down anaerobically. Sometimes they would forget and I would fetch them from the bins, the bags already boiling with the energy of decomposing grass clippings.

I offered to take on the garden myself, told the residents they could save service charge money by just buying me a rake and a lawnmower. They didn't trust me, didn't think I'd keep it up; they were worried about mess and rats. Each tree that died (due to a lack of water) was replaced with a palm tree, each dandelion drowned in poison. I tried to focus on my small successes: blue and great tits, blackbirds, woodpigeons, bee-friendly flowers. But so many things were destroyed: crocuses planted in the lawn mown just before they flowered; herbaceous perennials trimmed back where they would die, weedkiller, those bloody purple pansies. We don't want it to look like a jungle, they would say. We can't have that.

I bump into one of the committee members as I wait to be let into the complex, and she asks me if I'm living in the country now. You like nature, don't you, she said. She once told me she hated it, loved the bright lights of the city. I take a walk around the communal garden. A new building has shot up behind it and the lawn bears the muddy marks of no longer getting enough sun. The green roofs look good, they've had their annual autumn trim, but my ox-eye daisies look like they've been nuked and the bee-friendly plants have mostly shrivelled and died. Another silver birch has gone and in its wake is a palm tree, while a straight line of brightly coloured pansies glows like radioactive waste in the foreground. Fruit from my ornamental cherry spills onto the poisoned land; I hope/fear the woodpigeons are making the most of it.

I leave my former home and walk through the park, past gardens I used to love that have now been paved over. A car park butts up against severed back yards, weed-suppressant membrane and stones cover earth along its edge. The garden where a large, gnarled elder grew is now beneath paving stones, and there's another further up, where a London plane tree grew. I watch a digger carve up a patch of brownfield land, where buddleia and willowherb served butterflies in summer, where foxes played beyond Do Not Enter gates, where house sparrows nested. Luxury flats soon, no doubt. Maybe a sprayed-green communal garden with gaudy purple pansies if they're lucky.

What are we doing to our land? I think of everything I've achieved in the postage-stamp garden I brought back from beneath paving stones. That could be any patch of land, anywhere. There could be a network of native shrubs and trees, long grass, wildflowers, in gardens and balconies up and down the country. Some of our wildlife could survive here. It's not too late.

✿❀✿

In Hove the blackbird sits on the trellis, shuffled up, grumpy-looking, annoyed that I've disturbed him. He's not singing – he does that from the top of the Leyland cypress three doors down – but perhaps he's roosting here, or trying to. It seems such an exposed spot for a night-time roost. But as twilight fades to black he will gradually blend in with his surroundings, I suppose. Perhaps the house sparrows don't like to share the holly.

The blackbird is new; a young gun, I think, a male born last year trying out a new territory. My camera traps catch him poking around on the ground, and he serenades me at dusk and dawn, his flutey whistles and rasps the town crier singing *March* for all to hear. Mostly he's a chorus of one, occasionally joined by chirping sparrows and clacking gulls. I wait, patiently, for the day I'm woken by the dunnock and the blackbird together – what an event that will be.

He doesn't have a mate yet and I don't hold much hope for him finding one. I've seen it before, in London, a young male establishing my garden as part of his territory but failing to woo a female. It was winter, a female started coming in and then this male, full of hot air he was, spent the whole time chasing the female out of the garden, competing with her for food. They were so horrible to each other that I named them Sid and Nancy. And then, just as spring began to break the female disappeared – a winter migrant, of course. Here perhaps from Scandinavia or Russia for our milder weather, she would have set up home in my garden, eaten my seeds and berries, bathed in my bath, and then flown home again in search of a mate who was nice to her.

When Nancy left, Sid became bolder and more full of hot air. His territory seemed to take in the silver birch on the other side of the cycle path at the end of the garden plus the bit of train track that ran between two stations. He would serenade us with loud squawks from a TV aerial on top of a block of flats on the adjacent road, chase off rival males with as much gusto as he denied food to Nancy. Once, I stood at the back window and watched two male blackbirds compete for a female.

I hoped so much that the more successful male was Sid, but it wasn't to be; his rival and the female flew off into the sunset together, leaving Sid to lick his wounds in my honeysuckle.

He spent that summer in my London garden, ferreting around my pots for vine weevil grubs, practising his song from the corner behind the cherry tree. He sang from his rooftop TV aerial and scanned his kingdom from the lip of the railway bridge. He was a feisty bird and a terror, but I was so in love with him.

Like Nancy, one day he disappeared. I wondered if he'd moved territories or found a mate elsewhere but he most likely died. His patch took in a road, a cycle path and a railway – plenty to be lost in. But he could also have come to deliberate harm; I heard neighbours complaining that he woke them at 4 a.m. every day. Perhaps they took an air rifle to him.

So this new blackbird in Hove, this young gun, is welcome to sit grumpily on my fence, serenade me with spring, knock all of the soil out of my pots in search for grubs if he wants to. I'm glad to have another bird in the garden, I'm glad it's part of a wider territory. But I won't hold out hope, yet, for the patter of tiny blackbird feet.

It's lovely, though, to have birds in the garden, to be woken by a blackbird – there wasn't one last year. Really, now, it's the garden's first spring. Last March it was still mud and stones, the climbing roses still their bare-root selves, the decking still in piles, the rubbish I'd not taken through the flat. Now when I lean against the wall and drink tea, or crouch down in my corner, I feel the garden has grown into something independent of me. It has a life and rules of its own – leaves and shoots poking through the soil, plants seeded in from elsewhere. I have given it life and it has taken it with relish. It can only grow and get better.

The winter tits and goldfinches have gone, dispersed back to breeding territories, no doubt. The robin, too, which I've seen for two consecutive winters but never in spring and summer. The camera traps are still picking up foxes, which is wonderful, and my dunnock, of course, which continues to hang out around the edges. Leaves are unfurling, buds are bursting.

✿❀✿

Peregrine falcons are nesting on the seafront. They're at Sussex Heights, a twenty-four-storey block of flats between Regency Square and Churchill shopping centre. They've been nesting there since 1998, only once, in 2002, choosing the West Pier instead.

Sussex Heights is 102 metres high and has some of the best views of Brighton and the sea. Celebrities have lived here. Some of its flats are owned by DJs, their parties renowned across the city. I've never been to a Sussex Heights party but I know people who have. Wild, they say, crazy. Best parties in Brighton. Peregrines on the roof.

They're beautiful birds. Large, blue-grey on top with their breast and the underside of their wings like the perfect knitted jumper – thick cream collar with cream and white bars beneath, giving way to longer cream and grey striped wing and tail feathers, matching cream and grey striped trousers, bright orange feet. The face is black-grey with a moustache that bleeds into the cream collar, the eyes huge and searching, the hooked beak black and yellow. And the shape of them, oh the shape of them. Angled wings, splayed tail, the way, when they hover over prey, their feet hang like roosting bats.

The fastest animal in the world, the peregrine falcon can reach speeds of 200 miles per hour, hurtling through the sky in a terrifying swoop to ambush their prey. I've never seen one do this but I saw one take on a buzzard once and the buzzard lost. Traditionally they nest in more rural locations, on cliffs and mountains near a river or coast. They eat medium-sized birds such as pigeons, ducks and grouse. And therein lies their problem: in the last century they have been persecuted because they ate into the profits of grouse shoots or pigeon races. They had their eggs stolen for collecting or falconry. Pesticides in the food chain added to their woe and in the 1960s, the same decade their Brighton home was built, the peregrine falcon nearly died out. Yet around that time they started adapting their habitat, as if they knew that rural life wouldn't get them anywhere. They took to nesting in more urban areas, on church and cathedral spires, the first of London's skyscrapers. There are thirty nesting pairs in London now, others in cities across the country. And here, in a bespoke nest box on a 1960s block of party flats on Brighton seafront.

They're safer here. Safer in our tower blocks and cathedral spires, safer where the parties are, where they mostly go unnoticed but where those who notice them welcome them because they're beautiful and they eat pigeons. Outside the city they still face persecution from grouse-moor owners and pigeon fanciers. Their eggs are still at risk from collectors and falconers. Those that fledge from our cities come unstuck if they venture back into rural areas to breed. They're evolving to stay in our cities. They belong in our cities. It's thought there are more than 1,400 breeding pairs in the UK now, the highest numbers for fifty years.

Their nest box is lined with gravel and set up with cameras so people like me can watch them while we work and write our books. Nothing much is happening yet, no eggs but a nesting pair. They bring each other pigeons. I take binoculars to the seafront to watch them and take detours on my runs and glance up. But there's nothing to see, not yet, they're just beginning. Life is just beginning.

✿❀✿

I dream I'm in bed at Driftwood, in 1989, the birds singing so loudly I'm terrified they'll come in. They must be on the gutter, these birds, singing away in the darkness, on the roof or in the eaves. They're *so loud*. I feel like they're in the room. They're shouting, not singing. Shouting and threatening to come in, to fly around my room like the starling that came down the chimney a few weeks ago, the starling I thought was a crow. I like birds, but not when they're flapping around my bedroom. Not when I'm eight years old.

Granny comes in. Are you awake? Granny! She perches on the end of the bed, her green dressing-gown wrapped around her. Curlers in. I don't recognise her like this. I've seen her only fully clothed, Sunday dress or old blouse and trousers suitable for gardening. Pinny for cooking, flour or grease streaked across it. She looks old.

Can you hear them? The birds? I tell her I don't like it. I'm tired and they woke me up. Are they coming in, I ask. No, darling, she says. They're just singing. They sing a lot in spring. The males defend their territory, their nest with its female and eggs. Every tweet is a rallying cry that they're still alive and they will not be cuckolded. Here I am Here I am Here I am Here I am. That's all they're doing, there's no need to worry. She speaks with a lilt, like a bird. I don't like it, I repeat. Can you make it stop? She asks if I would like a cup of tea and I say yes, even though it's dark and no one should drink tea in the dark unless you're getting up early to go on holiday and this doesn't feel like a holiday. I ask her not to be too long. Don't leave me with the birds, Granny.

She returns with tea. Opens the curtains and we watch the sun rise behind the houses opposite, a fire above the chimneys. The blackbird sings first, she tells me, then the robin and then the wren – there, do you hear it? The wren is deafening, terrifying. I never want to see a wren. And there, she says, the chaffinch, and can you hear the chiffchaff now? It comes from Africa, like the swallows and the cuckoo. It flies here every spring, to have a family. I sip tea as my head fills with song and knowledge that I don't quite believe, wishing I could just go to sleep again, wishing this wasn't happening. My granny knows the birds because she's old. Old people like birds. I don't know why she's bothering to teach me.

The great tits start as traffic begins to roar, and the magic, or the madness, is lost. Granny lets me sleep in for a while as she has a bath, takes her curlers out. We eat breakfast with Sheba, watch sparrows and tits battle for bacon rind and stale bread on the bird table. I watch them, silently. Granny asks me if I'm OK. I'm OK. Do you want to feed the birds? No, I don't think so. What shall we do today? I don't know, Granny. I look down at my lap. She clears the breakfast things and we walk the dog along the lanes, me suspicious of every bird in every tree. Later she calls Mum, who comes to pick me up two days earlier than planned.

I wake up in a panic. It's dark and silent still. The memory of the dream crashes into me. Is that what happened? Did I go home early? It's 5 a.m. The Sun rises at 6.45. I lie in bed, staring towards the ceiling. I can hear the fridge kick in from the kitchen. Should I get up? Can I? Yes. I can feel her with me, sitting on the edge of the bed, green dressing-gown over her nightie, curlers in.

Did she even wear curlers?

I peel back the duvet. Stumble into the kitchen and boil the kettle, fill a pot with tea. I dress, tracksuit bottoms over pyjamas, a big thick jumper.

Nautical twilight, two hours earlier than when I last saw it, last solstice, to watch the starlings. It hurts a bit. It's freezing. I grab my sleeping bag and open the back door as quietly as I can, step outside into a perfect world that doesn't belong to me. There are no neighbours smoking cigarettes or twitching curtains, no one to avoid saying hello to, to sulk at for watching me; animals have the upper hand now. The garden at night is a different beast.

Ssssh. I drop my sleeping bag on the deck chair and return for the pot of tea, some milk and a mug, plus a pillow so I can lie back, half-sleeping, listening.

There's a herring gull on the roof of the house opposite, where a nest usually sits out of the way between two chimney tops. There's always another nest among the chimneys of the house next door; probably one on the roof of every house on every street in Brighton and Hove, but it's these two I see every day, from late spring through summer. I wonder if they've started nesting. I've watched them raise their babies, seen them grow and become bolder, bite their parents' legs when they're hungry, peck at slate tiles when they're bored. The parents often ignore them, turning the other way, staring moodily at the horizon. They make me laugh. This adult, a future parent no doubt, stands alone, tutting into the twilight, oblivious to me and my nightwear, so out of place in this netherworld hitherto absent of humans.

I tune into birdsong. The blackbird has already started and there's a robin, somewhere in the distance. The birds with the biggest eyes rise first, those that can see better in the half-light, the early birds that get the worms. Here I am Here I am Here I am Here I am. That's all they're saying, says the ghost of Granny ringing in my ears. Nothing to worry about. I climb into my sleeping bag and draw it up around me, pour tea, lean back against the pillow. A gentle breeze washes over me. The darkness is lifting slightly. I sip tea and close my eyes.

For a while I manage to block out the sound of the tutting gull with the flutey song of the blackbird, who sings, triumphantly, from the Leyland cypress tree three doors down. The blackbird is the friend you hate, the clown. One minute the Big-I-Am and the next he's crying into his cornflakes. Loud, bolshie, good-looking, paranoid. Silly thing.

A wren starts up, somewhere, which I love so much now but was so scared of as a child. I wish I could tell Granny. The wren is your little friend everyone pokes fun at but who doesn't care because she's fierce. Who's dancing on the podium with her belly hanging out at 3 a.m., spilling drinks over her mates. The one you can't see in a crowded room but you can hear because she has the most infectious laugh you've ever come across. That's the wren, Granny. I wonder where it's singing from and if it's ever come into the garden. I caught a glimpse of one in the smoke bush once. Could this be that one?

What's next? Blue tit. Like a wren that can't be arsed. Then my lovely dunnock, who has something to tell you but can't quite get it out. Breathe, dunnock, breathe, get it out. It's not the most exhilarating dawn chorus but it's not bad for a depleted city habitat mostly drowned in cement.

I stay for ages, refilling my mug with tea from the pot. It's cosy here, special. The birds sing almost in harmony, competing with the gulls, which now circle and cackle and call above me. I stay until it's fully light. I see shadows hulking across lit bathroom windows, curtains resentfully pulled open. Friday. Everyone's knackered, no one wants to go to work. The noise from the road increases as the dawn chorus diminishes, and the magic fades for another day. I watch house sparrows land in little squares of trellis, hear the first bumblebees start up, the low buzz of queens as they zone in on this patch of potential nest or that. I gather my things, return indoors to deadlines and bills: another world, the real world. I say goodbye to the gulls and the blackbird, the robin and the dunnock. I walk back to the flat with a laugh in my eyes. What a way to start the day.

PART TWO

A phoenix

Spring

It's the robin that sings first. A full hour before the others. Then the blackbird starts up in a huff and a puff, like it's been disturbed somehow, like its alarm has gone off and it's pressed the snooze button and rolled over back to sleep. OK, I'm here, says the blackbird, just give me five more minutes. And then the wren, the full song but with hardly any puff. Five more minutes for me too, says the wren. It's the school register and some of the students at the back aren't paying attention. Robin goody-two-shoes is making them all look bad.

Silence for a bit. In the distance then more robins and other blackbirds, less sleepy ones. Then the great tit and woodpigeon, chaffinch. The wren starts up, better this time, louder, more air in his lungs. The blackbird is still snoring. Five more minutes, he mumbles.

I lie awake, listening to birds through the open window. I'm not so scared of them now, Granny. Of these birds, these Birmingham birds. Mum's birds.

It had been such a happy day. Despite the dream about Granny, and the anxiety of those initial, waking moments, I had a special time listening to birds in my hidey hole, wrapped in a sleeping bag, hidden from view in the grainy light. I had planned to spend the day working but it wasn't to be – Helen called, some emergency with the twins and would I look after one of them, little Hester? I arrived at breakfast time, where I managed to feed and dress a tiny human and pack her things for the day she would spend with 'Kate Flower'. I strapped her into her pushchair and walked her back to my flat, where we played hide and seek, ate yoghurt and watched, possibly, one too many episodes of *Peppa Pig*. We looked around the garden but there's so little to see, still, at this time of year. Instead, I took her to a friend's allotment, where we found newts, a frog and a slow worm. Hester helped with some watering.

Later, after clearing up the toddler tornado that had stormed through my flat, Trudi and I made pizza, topped with the first, precious salad leaves of the year. Friday night. We drank one beer each. Luckily, just one beer.

Ellie called at 10.30 p.m. I answered, as I always do when family members call at the wrong time of day: What's wrong? I knew, instantly. Her voice had lost its bounce. It's Mum, she said.

She'd had a brain haemorrhage. A grade-4 subarachnoid bleed. Not Friday but the day before, Thursday. She'd just finished tutoring, had a bite to eat, that borscht I found in the fridge two days later, I think. She, Anna and Anna's boyfriend Ryan were getting ready to go and meet Pete and Ellie for a drink. She sat down suddenly, her head in her hands, said she felt funny. She became hot and Anna took her outside to cool down. Then she said she felt sick and so she ran, blood haemorrhaging into her brain, up the stairs to evacuate every last drop of everything that was in her.

Everyone thought it was food poisoning. The borscht or the sauerkraut maybe. She was up all night with it, blood spilling unnoticed, water on the brain. Eventually she stopped and she crawled into bed, to sleep. No one could move her. She slept like that for hours. Until the following afternoon, around the time I was getting ready to drop little Hester back to Helen's, when, finally, an ambulance was called.

From A&E she was transferred in another ambulance to Birmingham's Queen Elizabeth Hospital, to one of the best neurology departments in the country. It's where they send the war veterans, where little Malala was treated for her gun wound. The surgeons fitted a stent to ease the pressure from her brain, put her in a coma. By then I was tearing up the M40, tearing up everything.

Trudi and I arrive at Mum's, let ourselves in using the key hidden in the garden. Everything is as it always is, except there are sheets in the washing machine and stains on the carpet. I hang the sheets on the clothes horse, tidy up a bit, clean the bathroom. There's Mum everywhere: the book she's reading, the crossword she didn't finish. I find what looks like a shopping list written on the back of an envelope. I trace my finger over her words, written in green pen. Ingredients for a meal she would cook for someone with an allergy, maybe. No nuts!!! it says, underlined twice, and with three exclamation marks. I take the list, sneak it into my pocket. It makes me feel close to her, somehow.

Back from hospital at 4.30 a.m., Pete and Ellie sit with me around the coffee table. She's stable, they tell me. There's nothing else to say. Visiting hours don't start until 11 a.m. and so we go to bed, try to sleep. I lie awake, listening to the dawn chorus, as I had done just twenty-four hours before.

Ellie greets me at the door to the ward. Tells me to take deep breaths, prepare myself for what I'm about to see. I shrug her off. I can handle it, I tell her. I walk through, rub my hands in antibiotic gel, greet the nurses, turn the corner, see Mum, collapse on the floor. She's all swollen up. Huge, like a seal. Partially shaved head, tubes from her skull, her neck, both her wrists. She's not alive in the truest sense. The life-support machine is keeping her breathing, monitors are displaying her blood pressure and heart rate. I feel sick.

The surgeons decide there's no need to sedate her as much and so she's brought round to a level of consciousness just below awake. It's awful. Her eyes are shut and she doesn't speak, but she claws around, tries to take the tubes out, tries to sit up. Fighting back tears, I soothe her. Kate's here. Ssssh. It's OK, Mum. Mum? Kate's here. A mother and child in reverse. She hears everything.

We leave at 8 p.m. Return to her home, to the borscht in the fridge, the stains on the carpet. Again I can't sleep, the birds and the church bell marking each hour. Robin at four o'clock, blackbird and wren at five, greenfinch and blue tit at six. At 10.30 I walk around the lake, as she does every morning, while surgeons send a camera and apparatus into her groin and up her arteries to fix the burst aneurysm in her brain. I take the binoculars but I can't concentrate.

I deadhead her daffodils, use sticks to prop up her flower-heavy hyacinths. I water her pots, prune her espaliers. I take photos of the garden to show her: pulsatilla, pear blossom, narcissus, the cat. I don't know if she will see them.

The operation is a success and we sail in to greet her but she looks worse than ever. Swollen neck, more tubes, beeping machines. We cry as the nurses tell us how pleased the surgeons are. Every other person in the ward is in a coma. There's no window but it's a beautiful spring day. Chiffchaffs and coal tits sing from hospital trees.

They start weaning her off the drugs at 2 p.m. to bring her around by three. We take it in turns, two at a time, only ever two at a time. Nothing for ages. Be patient, say the nurses, she's been sedated for two days. Eventually she stirs, first her legs and then her arms. Mum? Mummy? She opens her eyes and turns to look at me. Mum? Can you hear me? Mum? I love you, Mummy. I love you. She doesn't speak or smile but I see love in her eyes. A smile in her eyes.

How's she doing? I ask all the questions. The nurse shows me scans of her brain, draws me diagrams, explains the risk of vaso-spasms, which can cause a stroke. I see her aneurysm and a map of the blood that burst from it. I wonder, in all that mass of brain and blood, where Shakespeare is – she loves Shakespeare. The nurse checks her eyes every hour, measures her blood pressure, her temperature. The machine beeps and my stomach churns. What's it doing, what's happening? These machines have a life of their own, she says.

Let her sleep, say the doctors. She doesn't need you now. Let her sleep.

I have no choice but to come home again.

The garden is parched and I water it. I hold the hose here for a few minutes, there for a few minutes. Everything is greening up without me. Clematis weaving into trellis, roses coming into leaf. If I stand here for long enough the plants will knit around me, above me and over me. Envelop me, hide me. No one will find me and yet everyone will know where I am. Like a long-forgotten bicycle trapped within a tree trunk. Loved, hidden, enveloped, smothered. Killed. By plants.

One of the sealed mud cells in the fancy bee hotel has a hole in it. A bee woke up and emerged from its muddy prison while I sat in Critical Care. Was it when she was being operated on or when she briefly woke up? When she clawed at her tubes or when she started drifting into a coma? When I took the borscht out of the fridge or moved furniture for the carpet cleaner? I tease out the internal wooden block and see more bees emerged from their cocoons, waiting for others in front of them to wake so they can fly out. A queue of bees. I unscrew the Perspex and set them free: two males and a female. The closed cocoons I lift out, gently, with the end of a teaspoon and transfer into the release chamber – I should have done this in autumn, a teaspoon is a better tool than a flat-headed screwdriver. One cocoon is half-open and I stand with it in my hand, a new life hatching on me, wiping its eyes on me, flying off me.

I don't know what to do. I feel so helpless here but at Mum's I'm just in the way, competing with everyone else's grief. But what am I supposed to do? I can't do anything.

Every morning I call the hospital. The nurse on night shift finishes at 8 a.m. so if we speak before then they can tell me how she spent the night. I call at 5.30, 6.00; sometimes earlier, sometimes I call at 3.00. How's she doing? She's stable. How's her blood pressure? It's fine. How long will she be like this? We can't predict that. What's she doing now? She's sleeping. Look, there's not much more we can tell you. She's sleeping and she just needs to sleep. I feel bad for wasting nurses' time when they're already so overworked and I try to stop calling but I can't. I want to be with Mum. I want to be the woman opposite Mum who sits at the end of her partner's bed all day even though he has no idea she's there. I've never been in Critical Care before and I never want to again. People no longer people but machines. Nurses stationed at the end of each bed tasked with keeping one person alive, one machine from beeping. Purgatory – most people are in a coma or nearly. Opposite Mum are two men in a coma. To the left of them is a woman who looks like she'll never recover from whatever horrendous thing happened to her. She stares, blankly, at the wall, cabbage-like. Is that Mum's future? Mum sleeps. I'm better off here.

But what am I supposed to do? When I cross the road I'm nearly run over; when I ride my bike I fall off. I've already scraped the car. So I stand, at the top of the steps, drinking tea or crouched down in my little hidey hole, watching the garden.

The sparrows, at least, seem happy. They're nesting somewhere but I don't know where, in the eaves of these neglected roofs, I think, not in the boxes I put up for them. When I'm still for long enough they come in, line themselves up on the trellis and sail down for a drink and a bath in the pond. I love them so much they make me cry. They take the nesting material I left for them in the robin nest box, pulling it out in great chunks and cheeping all the while. They venture into the border now. That's new. Hidden from view all of us.

The nurses tell me less and less and, eventually, I call the hospital and Ellie answers the phone. What's this? Mum's deteriorating. Should I come up? There's nothing you can do, we'll tell you when there's news. But wait, tell me more, tell me anything. Her arm is swollen. What? Her arm, it's perfectly normal, it's just swollen. But why? Because she's been lying in bed for a week. Is she still snoring? No, she's quiet now. Is she . . . stable? Yes. Is she . . . in a coma? A pause, as her voice falters. No. But? Look the doctors came round and she's not very responsive. Not very responsive. Her GCS has dropped a bit. Dropped a bit. To what? Four. From what? Nine, I think. You think? Yes. Nine? Yes. I repeat everything she says as if I've any clue what it means. I tell her I love her and to stay strong but I'm mad. I feel cut off, shut out.

I stay in my tiny flat with my tiny garden. Neighbours all around me. I sit outside as they prepare food, wash dishes, smoke out of the window. A knife scrapes food from a plate into a bin, a thumb clicks a lighter and is held, for a few seconds, beneath a cigarette, before being placed down as the first few puffs are inhaled. Lives carrying on while mine is on hold. I need space, I need to be alone but I never am. Yet there is earth, a spade. I can't sleep I can't think I can't eat I can't write. But I can dig. Each day my mum lies in her hospital bed, not living but not dying, I dig, transplant, weed. I don't eat. I drink.

I'm sitting in my hidey hole drinking beer. Monday morning, no one's here. I watch red mason bees go in and out of the bee hotels; they're using the fancy wooden ones now I finally removed the old box filled with hollow stems. Mid-spring and the birds are busy. Sparrows gather nest material, great tits burp from unseen trees. We're joined by a robin, who sits in on the trellis and sings of unresolved misery, of loss, maybe, of being misunderstood. All day. The thing is, says the robin. But you don't understand, says the robin. Let me tell you one more time, says the robin.

Mid-spring and there's plenty to do. I dig and weed and dig and weed. Prune a bit, plant a bit. Sometimes the grief catches me and I sink into my chair and wail with the robin.

The last time I spoke to her I was trying to get her off the phone. Imagine. If that's the last conversation I had with her then I don't know how I'll live. I was walking to the cinema and she called me. She was going on and on about this book I got her for Christmas. *Shakespeare's Gardens*. She was being sweet, really, but I don't know, I was busy I suppose, mind on other things. I got her the book as a stocking-filler. I thought she might leaf through it and keep it on the coffee table. But she read it from cover to cover and when she finished she wanted to redesign her garden as a nod to the great Bard himself. How I've walked into this one, I laughed. What are those trees, she asked, that grow as sort of lollipops? Lollipop trees, Mum? Great, I want a lollipop tree for my Shakespeare garden. What would you recommend? Saturday afternoon walking around Seven Dials. Bay? She doesn't want a bay. Box? Yew? Ooh no, not a yew, she says. Plenty of yew here already. And on it went. I'll have a think and get back to you, I say. And that was that. I went to the cinema and now she's nearly dead.

There's a blackbird singing full pelt in the street, the first of the day creeping above the curtains. On the laptop I watch peregrines. There are two eggs now, mottled red-brown. I see glimpses of them only, as the parents change shifts, one standing up and stretching its legs, the other rocking from side to side as it settles, shifting feathers out of the way so its brood patch – the area of bare skin on the breast – is in contact with the eggs to keep them warm. It's calming, somehow, to watch two giant birds raise young on top of a block of flats on a webcam on my laptop. Life goes on and here it is. 4.30 a.m. Nearly light. One of the birds shifts and stretches, lifts its great wings as if yawning. And then jumps up, drops down and out of sight. Breakfast time.

I look up GCS online. It stands for Glasgow Coma Scale. A normal waking person is a fifteen. After the op Mum was a nine but not any more. The scale starts at three; one and two must be *coma*. Mum's slipping into a coma and there's nothing I can do.

My head crashes back into the pillow and I lie, staring at the ceiling, blurred through tears. I have to go back up.

✿❀✿

A line of cars, a racing heart. It's been ten days and I'm less than a mile from her. Is she awake or dead? I pull up on a single yellow. I have to. I run out and all the way to the main entrance, up two flights of stairs. Past people shuffling on zimmer frames, pushing drips and oxygen tanks. Sorry, excuse me, sorry. At the entrance to the ward I ring the bell. Silence. There's another queue now, another line of anxious, desperate people begging to see their loved ones. I'm breathless but I don't care. People look at me. I ring and ring and ring and eventually they let me in.

Her bed's empty and I stand and flail about, still catching my breath and now in tears, as the nurses come to see me. Where's my mum? They must see this every day. Grown people reduced to babies crying for their parents. Who's your mum, darling? I tell them. It's OK, she's just been moved. Moved? Why? They mumble something about having an even mix of patients across a section of ward 'in case something happens'. I wonder where Mum sits on that scale of awfulness. Has she been moved because she's deteriorated? No one will say.

I'm shaking as the nurse takes me around the corner and points to a bed with a bruised and swollen lump from which a thousand wires and tubes connect to a beeping machine. Ellie on the chair next to her. Look who it is! Mum opens her eyes and moves her head to take me in: a sweaty, breathless, tearful thing aching for my mum. She greets me with a giant smile and a rusty *Herrrro!* My God.

The doctors and nurses have given her extra oxygen and raised her blood pressure so the blood pumps it faster around her body. It's had the desired effect: she's woken up. I can't hug her properly because of the wires and I'm worried I'll hurt her. I lean in to kiss her and take her hand. It's all I can do not to climb into bed with her. She looks at me with giant searching eyes. Hi, Mummy. Hi. Hello, Mummy. I take her hand and kiss it. I can tell she's using all her brain power to look at me. She looks and looks until she can't any more and she sinks back into the pillow and closes her eyes. Gone again. Ellie says she's been 'talking' for an hour only. What's she been saying? *Herro. Yes, no.* Early days then. Yes. Ellie leaves and I sit with Mum, desperate to wake her so I can have a piece of her. But I don't, I can't, she needs to sleep.

She's been moved to a bed by a window. There are workmen on the roof opposite and I wonder if they can see us. I wonder if they have any clue about what goes on in here, Mum sandwiched between someone about to have her life-support machine switched off and a man trying to pull his breathing tube out. I stroke her arm. There's a giant bruise around the blood pressure line that feeds into her wrist. She's been having problems with that, says the nurse. Has to be at the right angle. We keep taking it out and refitting it but it doesn't work, she's too fidgety. Fidgety's good though, right? He tells me her GCS has climbed a bit. He checks her responses as if bang on cue: shines a light in one eye and tries to do the other. She jams it shut. She's a fighter, this one, isn't she, he laughs, as he ignores her refusal to open her left eye and ticks the box saying she has anyway. He tries to get her to speak and stick her tongue out. She opens her eyes to look at him and then jams them shut again. I try now: Mum? Will you just do what the nurse asks? She ignores me, resolutely determined to sleep. GCS back down again. That's it? That's what measures consciousness? She's not slipping into a coma, she's just being bloody difficult.

I drive back to hers along roads hard-wired into my brain, along roads where I used to ride my bike, past the street lamps Mum attached Labour Party billboards to for the 1987 election, past my primary school, to the roundabout and – I stop. Park up for a second and drink in the view. They've had new windows put in. Tastefully done, Mum always said they were tasteful. New doors on the garage and side gate. Nice car on the driveway. Sticky rhododendrons coming into flower, that awful variegated holly still on the lawn. I look up at what was my bedroom window, imagine what lies beyond it now. The honeysuckle that grew up to the window is gone. I had the best room in the house, Mum said, because I got to smell honeysuckle flowers each morning in summer.

I've been sitting here long enough. I don't want to be seen. I pull out onto the road again, drive past the cherry trees in full, pink April blossom, as they have been every year for the last thirty years. The blossom that falls like autumn leaves. Do children kick it now, throwing it up in huge gusts? Do they gather the browning petals together like a snowball, throw them at their siblings? Past the big house where I used to sneak in and hide behind the conifers. Past the corner where those kids asked me if my bike came from Oxfam; where an old lady took me in one day and fed me ice cream.

Before I know what I'm doing I turn into the sports club, drive past the mere on the right, as far as I can go through to the end of the car park. Park up, sit in the car. Heart on fire.

It's different now. There's been an extension, I think, a second floor put in. There's a big gate now blocking my way to the fence I climbed all those years ago, to peer over into the garden. Peer at the manicured ornamental borders where my swing and the greenhouse once stood. The new fence that replaced the broken one that allowed me to sneak next door and gawp into their pond. The perfect lawn in place of the fire pit, the compost heap, the masses of cow parsley. The tennis court where runner beans and gooseberries once grew.

The ten-year-old me eyes the fence. It's 7 p.m., there are people about. People with tennis racquets and cricket bats, people heading to and from a gym which didn't used to be there. I get out of the car, stand on the kerb and peer as far along as I can but I'm not tall enough, not near enough, not young enough to get away with it – a 36-year-old would have far less bargaining power with the police for trespassing than a child. I return, reluctantly, to the car and drive off. But I'm not done. I drive around, past the house again and then down to reach the other entrance, past what used to be Mum's garden. I just want to see it. I just want to have one more look. I find the other entrance to the sports club and it's all barbed wire and CCTV now, a far cry from the 1980s where you could sneak in and pinch cricket and tennis balls to play with your mates. What is this obsession, these days, with shutting everything off?

At Mum's I pop my head through the door and greet three people in ruins. Hiya. Attempts at eating together have failed, we're better dealing with this on our own. None of us can really eat anyway. I change and head out for a run. Down the lane and around the lake, past the black swan and black-headed gulls, the fishermen with their spliffs and tins of lager, their bags of kit to catch fish they're not even there for. I stop and watch a pair of grebes rooting around in the depths. They seem ignorant of each other, indifferent, perhaps. They dive down, disappear, resurface elsewhere. Diving for what, I wonder, they don't surface with anything. They're far apart, these grebes fishing beneath the flight path. The cloud-spattered sky, pylon lines, planes and fishing rods reflected in the water. Oystercatchers peep-peep at each other, walking in circles as if on padded feet, one foot gently placed before another. A little gull soars and dips for insects. Around it other gulls gather into the sky, wheel around as the scene changes. The sun is setting. The fishermen are packing up. The grebes are diving down, resurfacing, diving down. Gradually, intentionally, they surface opposite each other. Oh hello, says the grebe to the grebe. They swim towards each other now, beaks slightly open and pointing downwards, eyes locked. They stop when they meet. Sit still for a moment. Then one flutters its head gently, coyly, as if to say, Look at me and my pretty grebe crest. The other follows suit. The first reaches behind itself as if to scratch an itch beneath a wing. The other follows suit. And then the dancing begins, tentatively at first, as if making allowances for early stumbles: a look to the left while the other looks to the right, a flutter of the head, a scratch of an itch. Look at me, no, look at me, you pretty thing, no, you are. Again and again in sequence. They swim away from each other, swim in tandem, swim towards each other. Dance again, part again, dance again. As the sun sets on the lake, as fishermen pack up their expensive gear, as people fly above us in giant tin cans to faraway places. As Mum lies in her hospital bed.

The sky burns. I leave the grebes to their seductive dancing and run another circuit of lake. Most of the fishermen have gone now, their litter discarded at numbered 'fishing' stations every few metres. I pick it up, zigzagging across the dimly lit path. Red and white clover, bird's foot trefoil and vetch are coming into flower as the grass thickens and greens, the trees in full leaf. How Mum loves this time of year. I press on, harder, blocking her out, working up a sweat. Each pounding of the path is a release, a break from the stress. If I run every day I will be OK. Somehow, by some means, I will continue to function, Mum will continue to breathe.

The path twists, putting the lake on my right and a small patch of trees and a stream on the left. It looks damp here, a carr. There's willow, alder and birch – trees typical of wet woodlands. I run fast alongside it to be quick. The sky is darkening now and the scene is changing again: cars pull up on the outskirts for other activities. Souped-up cars with big engines and loud stereos. I don't feel safe, suddenly. But there's something in the distance that makes me stop. Something like someone blowing on a siren whistle but not quite. I stand panting in the growing darkness and think of

Sheba straining on her lead, of Granny. The sound fades and I convince myself I imagined it. I start to run on again but then it comes back. Yes! It must have flown closer; I can hear it more clearly now. *Whoo-hoo. Whoo-hoo. Whoo-hoo.* The first of the year, the first in a few years actually, and the first, I realise, as tears well for the hundredth time today, that I've heard in this neck of the woods since Granny first taught me the cuckoo's call in 1989. It's all I can do not to sink to my knees and cry for the rest of time. What homecoming is this, that I nearly lose my mum and in the process remember every last thing else I've ever loved and lost?

She's getting better but 'better' is measured in horrifying quantities. She's not dead. She no longer has a burst aneurysm or an aneurysm at all. She's not slipping into a coma; there's no life-support machine, no ventilator, no extra oxygen, no help with breathing. She can cough; for a few days we were worried she couldn't. And she can regulate her own blood pressure. She still has that awful octopus of lines coming from her neck, the line in her wrist still surrounded by the biggest blue bruise. The oxygen thing on her finger. She keeps taking it off and I put it on my finger to stop the machines beeping and the nurses fussing. It reads 98, same as Mum's. The same oxygen coursing through our veins, the same blood.

She's a little old lady with a faraway stare. Someone I wasn't prepared to see for twenty years. Fragility happens to those lucky enough to live long but it's hard to see at 63. She's a hybrid of Grandad and Great-aunt Maureen, as she always knew she was, except she looks like they did when they got really old and frail. She looks like they did before they died. I'll never be able to erase this vision of her. If she gets through this I know what she'll look like in twenty years. I know what she'll look like when she's dying.

Slowly she turns to me. I take her hand and sink into her eyes. Hiya. She mumbles something in response, turns away to stare ahead again. It's as if she knows now that something terrible has happened. She knows she's broken. Her faraway stare is the stare of someone who can't believe this has happened to her, who's terrified she won't recover. I know her. And I can see through all the swelling and bruising, the beeping of machines and the care of the nurses: she's furious.

She opens her mouth to kiss me but it's like a fish kiss, a child's kiss, lips open, the kiss not fully formed. She closes her eyes again and the tears I'd been fighting back start to fall in great, heaving drops. I think of raindrops on leaves. The sound of summer, of replenishment. Stop crying, for fuck's sake. I look out of the window at the grey hospital roofs and think of my garden. Is it raining at home? I blow my nose and Mum opens her eyes again, catches my grief. She pulls a there-there face. A sorry-there's-nothing-I-can-do-about-this face, an I-barely-know-who-I-am face. I feel guilty. I don't want her to see me like this. I want to be strong for her, bear the cross for her, but I can't help making a little of this about me, about not knowing how much of a mum I have left. She's not dead but how alive is she, really?

The nurse asks for five minutes with Mum and I head out to the waiting room. I open the door to a family wailing in the corner. I close it again, head back into the corridor. I fire up the peregrines on my phone, watch the female sit on eggs. Four now. Nothing happens; I just want to watch them, just want to see someone else's reality while I lean against this hospital rail, antibiotic gel drying on my tightening skin, my barely conscious mum being rolled over and cleaned by a nurse with a giant heart. Four eggs. Four eggs in a nest on a tower block looking out to sea. Four eggs, two peregrines, my mum, this hospital and me.

The bell rings into the house. I see Ellie on the step, dressed in her blue and green checked dress with the white frilled bib, long hair draped over hunched shoulders, scowling face. Stand on the step, says Dad. Stand on the step and smile, no, smile! He takes a photo of a grimacing child: snap.

The man opens the door and I recognise him. He's older and rounder but otherwise the same. He welcomes me in through the door into the hallway. Into the living room. It's smaller than I remembered, of course. But everything is as it was. A new floor but . . . the old fireplace and shelves, ceiling beams, patio windows. My heart leaps into my throat. What am I doing?

It's been twenty-six years. Twenty-six years and I've spent three weeks driving around my old haunts to and from the hospital, and ended up here. It wasn't hard to arrange a visit. I searched online for the address and she runs a photography business out of here. The couple Mum and Dad sold the house to when they eventually divorced. I email her using the 'contact us' form on her website and she gets back to me straight away. A strange request but go on, she says.

We exchange pleasantries, my eyes on everything. I live in Brighton now, yes, write about gardening. Yes, I look like my mum. She's in hospital. Brain haemorrhage. Yes. Not really talking but moving limbs. It's difficult, yes. They've dug out photos, bless them. Photos from when they moved in, as it was when I left it. A far cry from the battered 1983 photo I keep in my bedside table.

She opens the back door through the new conservatory where the patio, hydrangea, lily of the valley and blue-tit box used to be. Out onto a long lawn with borders on either side. I drink in the view.

There's no tennis court.

I stand, speechless for a minute, as they chat about this and that. I look around, taking it all in. My eyes feast on a million things at once, brain recalculating and recalibrating. Things that were huge now appear small, things that were small now tower above me. There's forget-me-not, lily of the valley, rhododendron, but little sign of lemon balm or cow parsley. There's no long grass, brambles or ashy remains of fire. No vast, wild compost heap, no gooseberry bushes swamped through years of neglect. The canopy has closed above us; birds flit through unknown trees.

There's no tennis court.

There was a tree trunk here, he says, and a hundred-year-old greenhouse there, and in the undergrowth we found the remains of a swing.

A swing.

The ground was so uneven, we had it all levelled off, he says. Yes. I don't blame them, but it's harder to make dens on level ground. Harder to play King of the Castle or get a good aim when throwing mud pies at your sister. There's no vantage point, no little mounds you can claim as your own, dressed in shorts and a hand-me-down vest, oversized wellington boots, feather in makeshift bandana.

It's beautiful. Really, and it couldn't have been more nicely done. Primroses have sprung up in the lawn, travelling through it like a yellow brick road towards Emerald City.

That's an elm or an ash, apparently, he says, pointing at an elm that didn't used to be there. It's rare now due to some disease or other, he says. I pick off a leaf and show them the tell-tale sign of an elm leaf – the unsymmetrical base by the stalk. Elms break all the rules, I tell them. Most of them flower and fruit before a single leaf bud unfurls. I've seen quite a few lately, and I wonder if they're fighting back after Dutch elm disease brought them to the brink of extinction. Or if I just notice them more now I live in Brighton.

Do you want to take photos? I take photos. Do you have any more questions? I don't. We stand together on the levelled-off lawn, as I take in the forget-me-not and lily of the valley, the rhododendron and the still-there air-raid shelter, holding back thirty years of tears.

Did I remember it wrongly? Did I dream the loss? Did I peer over the wrong fence all those years ago or did I make the whole thing up? I swear I saw it, standing on the saddle of my bike leaned up against the fence. I swear I saw it.

But I didn't. I know that now. This thirty years of pain has been misremembered, wrong. There's no tennis court, no paving over of a wildness, of me. Instead there are self-sown forget-me-nots and primrose, a healthy elm – I couldn't wish for a happier outcome. Even the tree stump that we used to slide down as kids is still there. Flowers of my childhood still serving the descendants of the bees that flew around my six-year-old head.

Of all the gardens I have loved and lost, this one holds a piece of me. This, with my DNA from cut hair and skin from scabbed knees, dust of feathers collected to top mud pies, buried pet rabbits. We're in the soil and the leaves, the birds, the bees, little pieces of them and me. This garden is still mine.

It could yet be lost, of course. Like Mum's childhood garden that's now houses and driveways, like Tiny the pony's field that's now a postage-stamp garden. Neighbours have done it: knocked down a house and built three in its wake, removed a hedge to build an access road. A granny flat here, a garden office there, a fence, a wall, a paving stone, decking. My Spokey Dokeys clatter on wheels navigated through changing roads, a changing land. I still hear them.

✿❀✿

The doctors are playing with her blood pressure to keep her conscious but are mindful of the lasting effects of what they're doing. At 170/120 she's bright and perky but high as a kite. She takes her feeding tube and her catheter out, tries to remove the drain from her brain. She wants to get out of bed and she tries to swear at a nurse but she doesn't have the words. I tell her she's had a brain haemorrhage. She says, Heaven's Above and Bloody Hell. I tell her ten minutes later and she says, Blimey. I tell her ten minutes later and she starts to cry.

At 170/120 I have more than the mum I've had for two weeks but it's not particularly good for her. The doctors bring it down to 140/97 and she sleeps, wakes drowsily. I'm not sure she knows who we are. She takes my hand and kisses it, takes the hand of one of the nurses and kisses her. She sleeps, the only sign of her the constant fidget of her tiny feet.

Her nails need doing. I ask the nurse if she has clippers and she finds me a set: clippers, emery board and moisturiser. I feel guilty for not bringing Mum's. The NHS has enough to deal with without supplying the half-dead with manicure packs. I busy myself while Mum sleeps, cutting her nails onto a tissue, filing them smooth before coating them in moisturiser and then pushing her cuticles down.

I brush her hair now, which clings to her head, unwashed for nearly four weeks. The shaved patch is growing back, stubble around a red-raw hole the stent disappears into. I wonder how far down it goes, how it works. It's still dripping fluid into that container, behind her. It's pink, full of blood. Somehow the bleed got through to the other bit of her brain. Best to drain off as much as possible, the nurses say.

A woman shuffles into view and greets the nurses at the next bed. I just want to say goodbye, she says. The nurses are slow to act and the woman sits down and empties her grief onto the bed of her brain-dead daughter. Great heaving sobs as she tells her she'll never forget her, how she can't believe this is happening. Nurses jump to attention and draw curtains around the bed, contain the pain. She could be me, could be any of us. I can't look at Mum and I stare out of the window at the workmen instead, trying to block out the final goodbyes of a mother and lifeless daughter a metre away. Nurses in tears now. Mum opens her eyes and starts to sing.

✿❀✿

The sap is risen. The climbers are taking their time, the honeysuckle's still not reached the trellis, 'Jan Fopma' dying at the crown – clematis wilt, maybe, or a cat or fox sitting on its emerging stems. I empty a can of water on the soil, place my riddle over it and hope for the best. Above it, 'Frances E. Lester' is in fresh green leaf on one side, struggling to reach the trellis on the other. Be patient, say the plants and their reluctant stems. Everything will happen in time.

The garden no longer needs me. Apple blossom's been and gone; borage, honeywort in flower, kale gone to seed, purple-sprouting broccoli spoiled. At the back, 'Shropshire Lass' is a firework of thorny stems and shiny green leaves. I tie some of these into the wires on the broken fence panel. She will cover it yet, her leaves and flowers masking the broken grey. To her left is 'Bill Mackenzie', already halfway up his ladder, stems folding into trellis. Before long it will meet 'Shropshire Lass' and the two will grow and weave together, one flowering into the other.

The grass is growing, as I knew it would. New shoots greening the brown of the dead turf. Blades wave in the spring breeze, caterpillars hunker in the thatch. I set camera traps so I can see who my visitors are while I'm with Mum. Cats, mostly. But the odd fox and its babe, my house sparrows, my dunnock, my still-solitary blackbird. Herring gulls and their big white feathers, their big green shits. Collared doves, woodpigeons, feral pigeon, great tit, blue tit. As machines beep and responses are checked, a million worlds still turn.

A general ward now. Suddenly she looks like Mum again. Her hair greasy and shorn, smudged glasses. But Mum. Oh hello. Hello, sweetie, I'm just having a little nap. OK, Mum, I'm here. I smile, realising there's a little more of her than there was on my last visit. I acquaint myself with the new surroundings, introduce myself to the nurses, watch Mum sleep with still-fidgety feet. She's like a baby, all wrapped up in her swaddling of blue, sleeping peacefully now the wires aren't pulling her skin. She's eaten a jelly and is drinking through a straw. Feeding tube in only at night – progress finally. By day just the stent and the catheter remain, the latter the nurses have glued to her scalp to stop her trying to pull it out. The blankets are wrapped tightly around her and her little arms are drawn up to her chest. She keeps scratching her nose, touching her face, moving her little feet. She's a sea otter out on the ocean, on her back with her paws at her face, a little scratch here, a little wave there, bobbing in the deepest NHS blue.

I watch her sleep. There's a beauty to her I've never seen before, and I realise I've never thought of her as beautiful. I've never looked at her face and marvelled at her turquoise eyes and gently crooked nose. I've never looked at her in such detail. She's like a child again, my baby. I'm taking in every last piece of her, just in case. Just in case.

Mummy.

She opens her eyes as if too excited to sleep. She knows I'm here and she wants to see me. She takes my hand and mumbles deliriously, stuttering. The words have been knocked out of her. I tell her I can't understand her and she looks frustrated but she's too tired to try again. She drifts in and out of sleep, mutters about things that happened years ago. I can see every memory, thought and conversation storming through her brain as it starts to sift through the mess made by blood and fluid. Will she ever make sense again? She tells me she can't see. Darling, help me, she says. Help me. She's crying now, scared. It's her left eye, she tells me. She can't see out of her left eye. I ask her to cover her other eye and tell me what I'm wearing. A black jumper, she says, *shazooooziwus*. She starts to stutter and falls back onto her pillow. She can clearly see, but can she see clearly? It will be weeks before we know. I try to calm her, soothe her. I give her a cup of water, which she drinks through a straw. I wipe droplets from her chin, stroke her hair, balm her chapped lips. I call the nurse and ask her if she's mentioned her eye before. He says she hasn't and he makes a note of it. He asks her about it and she tells him it all started when she took Ellie to the garden centre. And I wonder when that was, and how her broken brain has linked it to her sight loss, and I lose her again for the rest of my visit while she chats about stuff that happened years ago in a language I don't understand.

Back at Mum's I look up her symptoms online while peregrines fuss over their young in the corner of the screen. One of the eggs has just hatched. The female takes a break from brooding to eat a pigeon the male has brought her, stepping aside to reveal the tiniest ball of nondescript fluff. I can't make it out, I can't see where the beak is, the feet are. Just a lump of pure white fluff, like cotton wool. The male sits on the nest and I see nothing again.

It's called expressive aphasia. These nonsense words that come out of her. Specifically she can understand what is said to her but can't make herself understood. Which must be awful for someone who's obsessed with language and Latin stems and taught English for forty years. What's the Latin stem of haemorrhage, Mum? *Hemo*, meaning blood. What about aphasia? *Phasia* is speech and *a* means not. She's speechless.

When blood lands at the front-left part of the brain it affects personality, behaviour and speech. At least her limbs work. We don't yet know if she will be able to walk but we don't see why she wouldn't, she's barely stopped moving her legs since she got in. But behaviour? Personality? That's a long road.

On clear dry nights over flat land, the air stills. It's called a temperature inversion, something to do with the earth losing the heat from the day. It creates a sort of echo chamber, an acoustic miracle, and the most perfect stage for a little brown bird to sing.

The nightingale has found a niche that suits it to sing at night, a funny quirk of evolution perhaps, when most other birds mainly sing at dawn. But it works. There's no one to compete with at night, no one to drown out the sound of you. And on a clear dry night over flat land, your song rises into the heavens so everyone for miles can hear you. On a clear dry night over flat land, if you're a nightingale, the world is your oyster.

It flies from tropical Africa and reaches our shores in April. Males set up a territory in the middle of a bush or a thicket and then sing each night to protect it. As they sing they might catch the attention of passing females, who fly overhead as they, too, return from Africa under the cover of darkness. And that is why the nightingale sings.

Scientists will tell you the part of the brain that controls singing is enlarged in a nightingale. They'll tell you the male has a repertoire of around 180 different song types, which he can pick and choose at will. That, if its song is broken down into syllables, the nightingale is some sort of god. The flutey blackbird 'makes do' with 108 syllables for his song, the skylark 341. The nightingale? 1,160.

His singing accomplishments also demonstrate how good a father he will be – the better the song the better the dad, the more musical notes the more often he will feed his hatchlings and the more he'll defend the territory against marauding males and predators. The nightingale song is nothing more than a manifesto: pick me and I'll do this.

Ignore all of that.

They used to be woodland birds, when woodland was coppiced for fuel and, as well as tall trees, in a wood you would find thick, low-lying scrub, patches of bare land, a bit of water and dense vegetation. You don't get that so much these days. So they've moved from woodland to areas of scrub on the field edge. Patches of land on public walkways and bridle paths. A bit of this, a bit of that: *voila!* Nightingales. You can't find them in Brighton, as far as I know, but there's a perfect patch of habitat just a twenty-minute drive along the A27.

I drive here on my way back from visiting Mum. On my way back from five days of watching her sleep and holding her hand, of trying to calm her, of trying to stop her crying. Of chopping food into manageable chunks and feeding her: just eat the chicken, Mum, just one more, please. On my way back from the M42, the M40, the M25, the M23.

It's a clear dry night over flat land and it's getting dark already. I park up at the side of the road. The long grass is dewy, despite the day, and my feet wet as they sink into it. I stretch my motorway legs and clamber over the stile onto the bridleway. Instantly miles away. To my right are thickets of scrub, to my left a little stream. I'm all alone. The dusk chorus is still apace, a song thrush belts out his tune from a tree while reed and Cetti's warblers call from the water. Wrens in the scrub. I walk for a while. The road is just a few metres from the stile but the remoteness, the darkness, the all-aloneness, is unnerving suddenly. The rusting ghost of the sun lingers in the distance and the night hangs and becomes closer. Birds shift among the trees. Where are the nightingales? I try to make out familiar shapes. The silhouette of a robin, unmistakable as it stands stock-still on its bandy stick legs, occasionally dipping down as if curtseying. It makes an alarm call and I wonder if it's me it doesn't like or if the encroaching night hides a darker force: a fox or badger perhaps. Rustling is unsettling. A sudden noise and I jump out of my skin: three turtle doves charge out from the scrub and disappear again. Again: what are they flying from? I pretend not to hear the shuffling behind me. The hedgehog or the . . . wolf. Eventually the blackness sets in and the dusk chorus fades. A tawny owl hoots in the distance.

Now the low burble of the nightingale rises.

It starts softly, hesitantly. Mosquitoes dance around me as the darkness takes the trees, the hedge, the architecture of the bridleway, and melts them into something 'other' and unknown. I can just about still see but I'm anxious now. I stand as quietly as I can, creep closer to the source. There are rich sounds, bubbly warbles, a few choppy calls, some gurgly frog noises and see-saws. But that's not what I'm here for. I'm here for the song that makes you want to melt, that makes you want to fall to earth and be swallowed whole. That makes you want to do air punches and cry, that makes you want to fly. That lifts you and floors you, fills you, reinforces you. And it comes, it comes, here it comes.

I close my eyes and wobble on two legs as something inside me lets go, as a chunk of hospital and motorway, wires and machines breaks off and flies away. Other males start up and there's now an orchestra of nightingales all along the bridleway. There's nothing now but me and them. The darkness, the stillness, the tears rolling down my laughing face. What is this? What is it that the nightingale does to the human heart?

We have evolved with it, certainly. Many of our ancestors will have grown up with it; those who lived or worked in woodland or its outskirts. The nightingale has sung for coppicers and charcoal burners, iron-ore miners, travellers. I imagine Robin Hood and his band of merry men being kept awake by singing nightingales, spring-born babes being breast-fed to the sound of nightingales, the old and sick dying to the sound of nightingales. Singing just from April to June it heralds summer, like the cuckoo. And the people would have taken its song as a sign of easier times ahead, of abundant harvests and good weather, of a heart-soaring celebration of still being alive after the cold, hard winter. Is that it? Is that all it is? Is the nightingale merely post-winter relief hard-wired into a haunting song?

Perhaps.

Nightingale song is a part of us, a part of what makes us. Yet, as we have moved from the trees to the cities, as we have changed the landscape and ourselves, we have lost that bit of us that connects us so completely with its song, and so too we have nearly lost the nightingale. It likes scrub and there's so little left of it. It should be so easy to just bring it back but we don't because so few of us connect with it now, so few of us are prepared to fight for it. It has sung through two World Wars, the Industrial Revolution, the beginning and end of whaling, of slavery. It's lived through it all, flying to and from Africa year in, year out, while we chipped away at its habitat. Between 1967 and 2007, the nightingale declined by 91 per cent according to the British Trust for Ornithology. It's suffered the second-biggest fall in numbers of UK breeding birds since records began. That's nine out of every ten nightingales vanished from our scrubland, nine out of every ten not calming anxious babes, waking highwaymen in ditches, soothing the dying or lifting the spirits of the hospital-weary. Nine out of every ten. They used to sing in gardens in London. Will they again? Most of us will never hear its song, most of us will never fight for it. But you would, if you heard it, you would. You'd die to protect the nightingale. To have that feeling rise up inside you, to be moved so completely. The very thing they inspire in us is what's needed to save them and that's the saddest thing. We lose them and we lose a central part of us. We already have.

If the dead could rise and make us see.

The darkness has taken everything. I think of folklore and fairy tales, of enchanted woods and sprites and goblins. Of this corner of Sussex and highwaymen and nightingales. Of what lies beneath these cushions of dewy grass, of If These Trees Could Speak. I wonder if I could sleep in this ditch as a nod to my ancestors and history and deep human connections with the natural world. But I'm terrified, suddenly, terrified of being alive. For the first time since I arrived I turn on my torch and scramble back, tripping over the cushions of grass, away from the rustling in the scrub. I reach the stile, the car, the streetlights, the safety and comfort of civilisation. I leave the nightingales for another time, another hospital visit, another year, and I'm grateful to have heard them this spring.

Mum calls me and it's almost her. A little confused, words still back to front, but she says things she used to say. She calls me darling, tells me she loves me, she just wanted to hear my voice. My heart melts and my brain struggles to understand her. But it's late, Mum. It's 10 p.m., the ward is full, I think of the old ladies she's disturbing. I ask if I can make a suggestion and she says yes, yes, and I say mightn't it be a good idea to go to sleep, with it being late and you being on a full ward? And she says, Hang on, I need to write this down because I forget everything these days. But she doesn't have a pen. She jumps out of bed and takes me, on the other end of the phone, through the ward, waking the sleeping old ladies: Do you have a pine? And I say pen, Mum, say pen, and she says what and asks another sleeping old lady for a pine, and the ward wakes and she descends into garbage and becomes agitated looking for this pen that she calls a pine, and she puts the phone down but still with me hanging on the other end, and heads out to the nurses' desk. I hear distant shouting. It's Mum. It's Mum shouting at the nurses. Mum who needs a pen to write down the instruction: Go to Sleep. Mum who can't be understood, with her lost mind and her half-speech, who is now ranting at nurses and little old ladies, who has clean forgotten she ever called me and why she got out of bed. I wait a few minutes as the shouting comes and goes, the moves of the nurses to restrain her. She called me because she wanted to hear my voice and I answered because I wanted to hear hers. But I've made everything worse as I knew I would. I put the phone down and cry.

I barely sleep, thinking of her in her new ward. In her, now, third hospital. Is she sleeping? Is she scared? Is she crying? Has she been sedated again? I almost hope she has. Knocked out, unable to feel what she can't quite understand. She's allowed to walk now the catheter and stent are out but she charges about on unsteady feet, crying and shouting. The first time she stood up she collapsed back in the chair again. The second time she made a few steps with one nurse on either side of her. The third time she ran and she hasn't stopped running. But she's unsteady. She shouldn't be running around like this. The nurses sedate her to keep her from charging about, to keep her safe. And to stop her waking the old ladies she shares the ward with.

We visit her and she doesn't remember anything from one minute to the next. I wear a shiny red coat like the one I had when I was three and she knows it, instantly. But she doesn't know my name. It's like seeing her through a steamed-up window. I wipe it and water streaks the pane, then steams up again. Her eyes, her crooked nose, there; gone. Each time she blinks it's a new day, a new visit. Where's Pete? Why isn't he here? He just left; he's been with you all day. She claims not to have seen him for weeks, her poor new husband keeping vigil by her bed for hours and hours at a time, unnoticed, unremembered. I thought seeing her in a coma was hard, thought the life-support machine and the wires and tubes and swollen arm and brain scans were hard. Now, suddenly, I'm dealing with the crazy, the fighter, the tearaway. The one the nurses are briefed on at each change of shift. The one who refuses food and medication, who has decided the consultant isn't kind and refuses even to look at him. The one who says No like an angry three-year-old. The one who sobs her way to the shower because she used to do this without two people helping her. The one who packs her bags and stands, defiantly, by the door until her husband comes to calm her. The one who has worked out how to use her phone and calls us every night to demand we get her out, crying and begging, Darling, why are you doing this to me?

She's aware enough now to tell us she hates it, to beg us to set her free. We tell her we're trying but it's difficult to know where to start, and besides, she's not ready. I call the hospital and ask when she's starting her speech-and-language therapy, if someone will be sent to look at her eye. As far as I can see now she's just being kept alive and safe. Three meals a day, very little medication. But she's miserable. They keep telling me she's on the waiting list for the rehab unit but they don't know when a bed will become available. I'm fighting to get her out of hospital but I don't know how any of us will cope if they send her home. Darling, why are you doing this to me? Darling, why are you doing this to me? Ringing around my head a thousand times a day.

In the woods I walk, slowly, tentatively, tuning into this song and that. There's a chaffinch and a blue tit overhead, a nuthatch in the distance. No people but me. Yet scrubby bits of bramble hide a thousand watching eyes. We are never alone, really.

It's late May but the bluebells are still in flower, rubbing shoulders with the last of the wild garlic and the first, bright-green unfurling fronds of bracken. The canopy is already starting to close above me but the leaves are still fresh and new. Sycamore flowers dangle from the heavens.

It's been seven weeks.

She hasn't called me today and I'm glad. She's someone else's problem today, I need to be in the woods today. Already I feel guilty.

I walk along my little wooded path, with each step a little more me, a little more free. A little less hospital than there was in the last step, a little less nondescript pie and mash, blood-thinning injections. Coal tits pass through branches like little gusts of wind and there's a low gentle hum of unseen bees. All around me, from the ground to the canopy, is green.

I climb up, off the path, find a little spot to sit down behind a tree, unseen. It's an oak tree, a great old thing with a big hole in its bottom. I'm free here. I'm an unfurling bracken frond, a gust of coal-tit wind. I lean back and rest my head against the trunk, let peace sink into me. The humming is louder here. A bumblebee nest in the bank perhaps, or honeybees in the canopy. I look up and down, try to locate the sound. I tune into different hums – it's not one thing but many, not one bee but an orchestra of flies, beetles and bees, zipping between flowers that meet a million different needs. I can't see anything. I scour the canopy but find only dangling tits picking insects from the outermost branches, a grey squirrel looking down at me. I search for too long and the sky moves above me, my stomach lurches and I feel the earth turning. I'm dizzy. I close my eyes and lose myself in the hum and thrum of the wood, the hum and thrum of decades and centuries past and future. The hum and thrum that soothed people who are dead now, will soothe people yet to live. Highwaymen and time-travellers, the living and the dying, the long gone, the not yet born. Ashes and dust. All around me, little unseen insects work from one bloom to the next. I want to curl into a ball and sleep here for ever.

I told Mum I was gay around the time Granny died. She's never forgiven me for it. All that going on and now this, she said. She told me I was going through a phase, said she didn't want a gay daughter. For years afterwards. And now she's in hospital and I'm feeding her, moisturising her hands and chapped lips, and I'm thinking of all the things I could have said to her, all the things she should have said to me. I could have timed it better, yes, but I was fourteen.

Yesterday she took my hand. She was crying, we both were. She told me she wanted to die. She told me to get on with my life, to stop wasting my time visiting her, sitting with her, feeding her. You should be out being brilliant, she said. She held my face and forced out the choked words, Don't Ever Think I Wasn't In Love With You Don't You Ever Think That. Because she thinks she's dying and she wanted to say the unsaid, because she nearly died and I need to hear it.

Hold her close. Tell her you love her. So says everyone who knows what it's like to lose their mum. Even if she can't hear you, just tell her, they say. Those who never got to say what they needed to say, who remind me that hope and love are the two greatest things, the only things, because they no longer have them. Hold her close. Just hold her close.

The things we push back and refuse to deal with, they all come to bite us in the end.

Three more days of hospital. Three more days of my mum not knowing my name. But she knows who I am. She asks me to tell her how Granny died, makes me relive my break-up three times. She sobs suddenly, irrationally, endlessly. I try to imagine having my memory wiped and then having it all forced back in a storm. A brief few weeks at peace because there's nothing there but then everything all at once. She has nightmares and I'm not surprised. How to even begin to get through this?

Sometimes it's the 1990s, sometimes the 1980s – she's rarely in the present. I sit with her as her brain churns through everything that's ever happened to her. She mumbles constantly but I can't understand her. Conversations she had twenty years ago, sifting through endless painful memories. She still tells me she wants to die, tells me to make plans. But she doesn't have the words or the memory to take the conversation further. Small mercies. She's not going to die, she's had her chance. She'll have, and is having, a recovery. No one knows how much of her will return but it's barely been eight weeks.

There's still no bed in the rehab unit but it's clear she can't stay here for long. It's not doing her any good. The nurses are trying their best but she's lost her mind and she's angry, confused. She blames us for her still being here, shouts and scowls at the doctors and nurses. At least if she were home she'd have her things, her memories. It might give her context, help her find her place in the world. But the thought of looking after her like this fills me with dread.

We spend my visits in the hospital garden, a hexagonal patch of green at the centre of the wards. She's not had sun for eight weeks and she soaks it up, dangerously refusing lotion or shade. Shall we sit out of the sun, Mum? No. You're looking a little pink, shall we go inside again? No. I try out speech and language techniques the therapist taught me, to help her relearn her language. Where's the bird bath? She points. Where's the aquilegia? She points. Where's the fuchsia, the buddleia, the lavender. Points. Nothing bloody wrong with her. What's my name? Jo. Oh. I show her photographs from home. The cat, the garden. Sometimes she refuses to look at them, sometimes she bursts into tears. The cat's missing her. Like us she's lost without her. Sleeps on her bed with her ears cocked, the faintest noise that might be Mum. We walk in and she turns away, disappointed.

While Mum sleeps I steal glimpses of the peregrines on my phone. The fourth egg has been pushed out of the nest but the lump of fluff has been joined by two others. I can make out all their faces, now, their barely open eyes, their beaks set like a downwards smile. The three of them shuffle into each other as if still in eggs, lifting their long necks only for food. Parents gently tear little scraps of pigeon for them to gobble down.

At first they come in twos and threes, distant shadows slowly wheeling. I watch them from my hidey hole, flat on my back in unmown grass: cloud-streaked blue dotted with circling, screaming home-again swifts. Finally, some Good News.

Like the cuckoo and nightingale they come from Africa in spring. They dive and tumble and scream us into summer, relentless and unyielding. Here only for three months but three months of fun, three months of tumbling through skies, racing around rooftops, three months of parties, of screaming.

I'm not sure where they nest; I've not seen them enter buildings. They choose holes in roofs and walls, ledges; few and far between these days as we renovate our homes, fill gaps to save energy. But in the neighbourhood, somewhere, in old schools, churches and homes, holes in the brickwork, neglected roofs, bespoke nest boxes if they're lucky. One day I'll live in a house with stairs and a roof and swift boxes, and I'll cluck around my nesting babes like an excited mother hen.

I shouldn't allow myself time for swifts but it's relaxing here, sinking into long grass, staring up at blue. Hospital-weary, my neck hurts, my body aches, my eyes are drowning in shadows. I close them and take in the sounds: bees, house sparrows, gulls, swifts. I'll get up in a minute. Ants furrow in the sward next to me; bees buzz in the blooms. There are butterflies and leafhoppers and centurion flies and things I can't identify. When I'm still, house sparrows line up, each one in its own square of bare trellis, waiting for the first one brave enough to sail into the pond, opening the floodgates for a flurry, a splash, of little brown birds having a wash.

Out of Critical Care but not yet fixed, the garden survives but grows the wrong way: the grass is long, wayward stems and gone-over flowers dance, unfettered, in the breeze. Plants grow into each other, over each other. Yet the trellis is still bare. But I feel safe and alive here, pleased to be here, the motorway etched onto my brain, the hospital tattooed on my skin. My hands flake from antibacterial gel. They need soil and thorn scratches, leaf juice. My nostrils need the nectar of a thousand blooms, my eyes and ears need these swifts.

Historically swifts aren't associated with fun. They've been called Devil's bitches, their screams the calls not of partying groups but of lost souls. How little our forefathers knew. They've got funny little faces and wide, gaping mouths. They fly so well they come to earth only to breed, but to a high vantage point rather than the ground, from where they can launch themselves back into the air, drop down and sail into the wind. They don't have the leg power to launch themselves from the earth. If they crash-land they can't fly off again, they require a human to cricket-bowl them back to the sky or drop them out of a first-floor window. I suppose, in less enlightened times, it would have been easy to mistrust a swift, a bird that turns up from nowhere in spring, flies around constantly but becomes grounded if it falls to earth, that screams in great piercing summer gangs before disappearing entirely. Before we knew about migration we thought they hibernated in mud. Imagine those large, sickle-shaped wings digging them out of the earth to lie, helpless and grounded while the others wheel about in the noisy sky. Devil's bitches indeed.

The swifts leave my patch of sky and I get up, weed out herb robert and sycamore samaras, which have seeded in every crack in the soil. I leave dandelions for the bees, manage the expectations of the green alkanet, make room for white deadnettle. I tidy, set this right and put that back. I've planted things too closely together but there's nothing I can do now. Lifting and dividing is an autumn job, for when the swifts have buried themselves back in the mud. I Chelsea-chop alternate plants so they flower later in the season, let some things bloom before others.

The honeysuckle still has not put on growth. It's yellowing at the leaves, stubbornly refusing to throw out a stem. It's barely any bigger than when I took it as a cutting from Mum's garden a year ago. Does it know she's in hospital? I water and feed it and make a note to do so weekly. It will be focusing its energy on root production, no doubt; next year it will climb. But it's painful, waiting like this, when I could have spent £15 on something that would be ten times the size now. We gardeners are such fools sometimes.

She's home and it's good for her. She's weak and confused but she's got some of her mind back, she's less mad. I drive her to visit Great-aunt Mary and she directs the way. We take a new route, a far cry from the same journey she's made weekly over the last thirty years. She tells me to go down this road or that, straight over at the roundabout. I'm learning to translate her new words: do you mean turn right? Yes. Do you mean take the middle lane? Yes. Eight weeks in hospital and she knows a better way. How? If the dead would return they could find their way home. Skeletons in the passenger seat – take this turn, get in that lane. Except you could go back only so far. Resurrect those for whom the land is unrecognisable and everyone is lost. For whom horses, not cars, got them from A to B, for whom the stars and moon, not street lights, guided their way. What would they say about what we have done? What would they do? The people and the bees too, the hedgehogs and birds, skeletons all. Look what you've done! Look what you've done! Too late. It's all too late.

We can reverse this. Some of it. We can rewild our parcels of land, grow things for bees, build new routes for hedgehogs. We can dig tiny ponds for dragonflies and frogs to breed in, we can plant trees that will serve birds for a hundred years. We can give back what we've taken away, rewild our land as well as ourselves. We could yet make mud pies and climb trees, we could yet gather moth cocoons. But how, says the child in the driveway playing with a toy car. How do we even begin?

I imagine the gardens along my road all linked together, all designed as mine, all wild. Each one full of house sparrows, bees. Each one with a bee block housing tiny solitary wasps, a pond with dragonfly larvae, a compost heap with centipedes and rove beetles. A fruit tree in every plot would make an orchard – enough for us to share with neighbours and birds. A long hedge at the back would make a route, a corridor that small mammals and amphibians could travel along. Imagine how many birds would nest in the hedge that replaced fence panels all the way from one end of the road to another? Each garden home to a dunnock and a blackbird that would serenade every last one of us each morning.

Between my flat and the Co-op there are three people who feed pigeons. The old lady at number 43 stands on the doorstep in her nicotine-stained dressing-gown, fag in drooped mouth, throwing stale bread into the road from her pocket. The man at number 92 half-heartedly arranges crumbs on his windowsill, the window open slightly so he might hear them coo-cooing from within his flat. Someone else feeds them in the most spectacular way. From the second floor they open a sash window as wide as it will go, where the birds queue for their daily treat. I imagine this person has a signal to call them with, an alarm going off, the dinging of a bell: dinner's ready. I walk past and see wiggling pigeon bums crowding the windowsill, jostling for space, as the shadow of a person behind them controls them like the conductor of an orchestra.

I watch the pigeons from my living-room window. They fly from one end of the street to another, a mass of grey bodies gliding against grey houses, from one pile of grey, stale bread to the next. I think about the starlings and the sparrows and the other birds that live on this street that my neighbours don't feed and I wonder why.

Others might think they're mad, might pour scorn on them. But these three people, who live between my flat and the Co-op, are nature lovers. I think about the green spaces they are exposed to, can get to. Few front gardens in my road have plants; these have long since been grubbed out to increase light to basement flats. The back gardens – who knows? It's easier to pave and deck than mow or tend herbaceous borders. The nearest park is a mile away. And who says if our three pigeon fanciers have access to them? Would they know what to do if they did? Do they know, even, that they like nature, are drawn to helpless creatures in the road? Do they see themselves as naturalists? I sometimes meet the old lady shuffling along the pavement. I doubt her journeys take her much further than the Co-op.

Nature is a healer. We know and have always known this. Regular access to green space reduces our reliance on antidepressants. Our blood pressure lowers, our heart rate falls. But you shouldn't need to escape your life to find it, it should be with you already, should be part of you. Yes, it's in the woods and the parks, the nature reserves, the mountains. But it's also on our grey streets. It's eating our food waste, nesting in our litter. It screams and shits, eats food, knocks over bins. It wakes us up. Foxes, seagulls, pigeons, rats. The 3 a.m. robin. Badgers and deer now roam the suburbs, soon they'll conquer our cities. If only we could see.

We're so far removed from the natural world that people who feed pigeons are more likely to be seen as mad or lonely than as nature lovers. We crave nature but we don't know how to look for it or get it back. We don't see the tools we have at our disposal, the space behind and in front of our houses. We walk the dog in the park while looking at our phone. Some of us feed the pigeons while everything else is dying.

So much of this could be fixed in our gardens.

I always smile and say hello to the old lady at number 43. Above her and next door, disgruntled neighbours have erected pigeon spikes to stop them hanging around after they've been fed. She doesn't seem to care. I see her defiant smirk as she scatters food over the pavement and road, as pigeons land on car roofs. Perhaps I should start with her. Have a word, donate a plant or two. Chuck a sparrow leaflet through the door – there's more that you can do than throw stale bread into the road. I can help you. Would you like that?

Summer

When I dream, I dream of Mum or the garden or other things like being lost at sea or hurtling through space, screaming. I wake up and feel like I've been in a fight. I try to stay sane by running. But sometimes I cry when I'm running and my breathing goes all funny and people think I'm having an asthma attack.

She, too, is still having nightmares. Each evening we take it in turns to sit with her as she tries to get off to sleep. She seems scared to let herself be taken, she resists it, despite her exhaustion. I lie in her bed watching peregrines on the laptop as she battles with sleep then falls too quickly and wakes with a jolt. Finally, I think she's gone and I close the laptop and suddenly she calls out. She turns, slowly, breathlessly, to see if I'm still there, a frightened child. She stutters her relief as she takes my hand: it's always nicer with a *fffffriend*. I kiss her and hold her and weep silently into her pillow.

The peregrine babies have names now: Diplo, Chaos and Xena. At least two are female. They're growing so big, the parents often leave them alone, huddled up together or playing tug-of-war with a pigeon bone or a sibling's leg. The nest is full of pigeon feathers, the gravel base barely visible.

Today I stop running because I see a coot out of the water and I am enchanted by its amazing foot. It's standing on one leg, revealing the perfect giant clown's shoe. A shoe made of pristine turtle skin. A shoe with incredible mottling and elegant sharp, pointed claws, a shoe not quite blue, not quite grey, not quite yellow. They're incredible, coots' feet, and mostly we don't see them because mostly coots are in the water or sitting on a nest. But this one is doing neither. It's standing on one leg, at the side of the water, being gawped at by a sweaty runner who hasn't realised she's stopped crying to look at a foot.

The garden is a coven of witches' fingers. Hunched-over spires of hooded bells twist and lean through the borders, each one ringing with the tinny sound of hidden bumblebee. We shouldn't touch them, or so everyone says, but I can't help it – there's nothing in this world to rival their softness. The feel of them transports me back to when they towered over me, when I would stand beneath one and look up into a world I shared only with bees, a world of fairy footsteps or haloed spots, guiding the way to the nectaries.

I'm deadheading. Cutting back here, sacrificing there. The compost heap is a mountain of colour and my foxgloves lie as hanged witches on top. As children we would wear the thimbles on our fingers. There were so many of them in June, the main border between the driveway and path full of them. For ages, uninspiring rosettes of slug-mangled, lizard-skin leaves, often yellowing at the tips. Then, suddenly, the most wondrous spires of deep purple, uncompromisingly announcing summer. Of all the flowers that take me back to childhood the foxglove is queen, but it's only when I touch the blooms or sit beneath one and look up into the internal workings of the thimbles that I feel it. Memory's a strange thing.

The origin of 'foxglove' is unclear but it may be a mutation of 'folksglove', referring to faerie folk. Scandinavian folklore tells of fairies teaching foxes to ring the little bells to warn others of approaching hunters. But also that the foxes' tiny feet were made quiet by wearing the thimble-shaped flowers as gloves, better allowing them to raid henhouses. Its botanical name, *digitalis*, comes from the Latin *digitanus*, meaning finger. Foxgloves have many common names: goblin gloves, dead men's bells, fairy's thimbles; but witches' fingers is my favourite.

If administered in high enough doses, foxglove leaves can decrease the heart rate, cause nausea, vomiting, diarrhoea and death – a botanist killed himself by eating foxglove leaves in 2005. And yet, if administered properly, they can help treat heart conditions. They can 'raise the dead and kill the living', so the old saying goes. Long associated with witches and witchcraft, foxgloves were used as a folk medicine for heart complaints, the knowledge passed down through generations of medicine women, mistrusted and tortured, burned or hanged. The knowledge survived Witch Fever and in the 1700s a doctor called William Withering learned of the remedy from 'an old woman in Shropshire'. Its main toxin, digoxin, is still used as a heart medicine today.

I grow foxgloves as a nod to my childhood and a nod to powerful women, witches. And for the wildlife, of course. I've seen only two types of bumblebee use foxgloves – the common carder, *Bombus pascuorum*, and the garden bumblebee, *Bombus hortorum*. Aggressive little wool carder bees, a solitary species that I've yet to attract to my bee hotel, too. Those with the longest tongues, necessary, I assume, for reaching the nectaries tucked up at the base of each thimble. If you sit by one you will see the bees zone in and out, working their way through the spire of blooms, travelling upwards as if climbing steps. Growing up to two metres and housing up to fifty individual flowers, each foxglove bears a lot of pollen and nectar in a small space. Good for small gardens, foxgloves.

It's not just bees that like it. The lesser yellow underwing, angle shades and setaceous Hebrew character moths use it as a caterpillar food plant. Biennial, there are so many in the garden now because the ones I planted last year didn't flower and then I planted more, just in case. Most are your bog-standard purple affair but there are some different ones, some cultivar or other, which are pretty and have pinker petals with a sort of internal magenta stain, rather than spots. I like them but I wonder if the bees do.

It's nice to be in the garden, which threatens once again to become a stranger now I'm so often elsewhere. Three hours on the motorway and I'm spending the last of the day here with the secateurs and still-at-work bees. House sparrows assemble in next door's holly. The blackbird on his Leyland cypress three doors down, a baby seagull each on its little bit of chimney. I trim the lawn with long-handled shears, snip rogue flowers here and there for the vase or compost, fish out blanket weed from the pond. The garden is mine again and I feel whole again. Only a few hours from the motorway and sadness of looking after Mum.

She's doing so well but she can't see it. She talks in riddles but mostly she tells us she's bored. And of course she is. She can't read, hold a conversation, go on a nice walk, do the crossword – all the things that make her Mum. She obsesses over tiny things – elements of her personality exaggerated. What time is it what are we doing next what's the plan where's Pete why don't you do this I think I might put my pyjamas on. It's 4pm, don't be putting your pyjamas on, the sun's shining, we're doing the crossword. But I'm bored I'm so bored, she says. I persuade her to come with me to walk around the lake. She doesn't want to, makes some excuse. But you're bored, Mum, come on! I know it will floor her but at least when she's falling asleep in the chair she'll forget how bored she is. Give it a chance, we can always turn back.

She reluctantly agrees. We walk to the car and I drive her down the road to the car park by the lake. Nice and easy. Out we get. Will we be long? No. Will we see anyone we know? No. But we might see great crested grebes or the black swan, and I heard a cuckoo a couple of weeks ago, Mum. Yes, all right, she says. Her anxiety is palpable and I wonder if it's a fear of being seen or if she just feels so awful that she can't bear to be away from home. It's heart-breaking bearing witness to this, from the woman who owns the dancefloor at family weddings, who plays table tennis on a Wednesday, who quotes Shakespeare as she skips down the stairs each morning, not caring for a *second* who she's waking up.

My hand on her shoulder. Come on, Mum.

We walk along the winding path overgrown with cow parsley, red clover and bird's foot trefoil. The first of the sloes, still small and green, tease us from branches above. It's quiet and calm at this time of day, most of the fishermen have packed up. Birds sail unseen among verdant trees. The last time she was here was March. Things were still brown then, just peeping through. Now everything is lush. We reach the lake and stop. Here we are, look, it's not so bad, is it? The lake stretches out before us, reflecting the pylon-striped sky. No, she says. No. She stands for a moment and I wonder what's going through her head as she looks at this view that, at one point, none of us thought she'd see again. I wonder if that's what she's thinking, if that's what this is – a reluctance to confront one's fears: I am here at the place I thought I'd never see again. I am here but not completely. I am here but broken, halved. I am half but this is still whole. We turn left as a cormorant smacks into the water for a fish. She stops again, surprised. It's OK, Mum. Then onwards, as baby great tits call from hidden nests, coal tits seesaw in the canopy, chiffchaffs flit half-heartedly. There's no cuckoo today. But there are grebes and I point them out to her as we stop and stand again. Oh yes, she says, *greebies*. She asks, constantly, how long we'll be and eventually I sit her down on the grass to stare at the water. She likes that, she says, but she's happy to turn back. For all her anxiety she walked around half the lake and we can tick off another thing she has done and Can Do. Back in the car to home and her chair, where she sits with a cup of tea and piece of cake and drifts off to sleep. I pack my things and drive back to Brighton, to my garden and the foxgloves. They can raise the dead and kill the living. What could they do for Mum?

Teenage peregrines poke their heads out of the nest as police sirens wail below. Xena and Diplo perch on the edge, pulling at their down feathers, desperate to be let go. Only Chaos still looks like a baby, stuck in the nest while his siblings explore. I feel sorry for him. I lie in bed with tea, watching them.

The garden is two. It's been two years since I exhumed it, brought it back from the dead. Mum's in recovery, the garden's in recovery. It's only me now who needs rehabilitating. Two months of stress and fear, of not eating, of too much drinking, of endless driving and of living out of a bag. I feel like it's I who have been exhumed. My neck aches with the weight of my head, every bone hurts. Like altitude sickness, when I stop I feel as though I'm glued to the bed, the chair. There's no energy left, no life. Every day I survive rather than thrive. Like Mum, like the garden. None of us is really doing that well.

I spend a rare couple of hours in the garden, inspecting this, cutting back that, weeding out this, moving that. I open the back door and the sparrows send alarm calls. They're in the borders, fishing out caterpillars and aphids for their young, I think. One sits on my wooden bee block and watches me while the others carry on until I move forwards and, *cheep!*, the lot of them disappears. How wonderful that they can hide here, finally.

It changes daily now. We've had rain and the space has gone mad. 'Bill Mackenzie' has claimed the trellis and is in full bloom, clashing wildly with 'Shropshire Lass' – I wasn't expecting them to flower at the same time. Self-seeded field poppies have turned up everywhere, competing with a sea of white and blue love-in-a-mist. Foxgloves tower, phlomis jostle, lychnis, cranesbills and cornflower flash, here and there, advertising their wares for a thousand buzzing bees. A small tortoiseshell butterfly rests in the lawn while a holly blue bounces through the borders. Wow.

I can barely get to the back now, to reach the bee hotels. I squeeze through anyway, lifting out the block to see nest cells made by eight red mason bees. Eggs and grubs abound, all looking good. There are no leafcutters yet. But soon. 'Frances E. Lester' is in bloom, her leaves as yet untouched but it's only June. There's plenty of time. My bee block has residents too. Although I've no idea who. Solitary wasps, perhaps. A lump of sawdust pokes through from one of the holes. I smooth it down and whatever is inside pokes it back out again. Whose garden is this? I sit and watch for a while but nothing emerges and my mystery guest remains just that.

It's so alive, suddenly. I fish weed out of the pond and a common darter larva jumps out onto the lawn. I ferret among the grass and find it, this squat, angry-looking thing, built for killing. What does it kill, I wonder, in my tiny water? Hard to imagine a full-grown dragonfly will emerge from here, mate and lay its own eggs in the water it came from. My two-year-old pond – I could cry with pride.

It's a sunny day and the garden is parched but it will rain later or so I'm told. I carve holes in the lawn and plant daisies, self-heal and bird's foot trefoil I half-inched from a bit of lawn on Nevill Road. Just red clover to find now and my list of lawn weeds is complete. Dandelions found their own way here but everything else had help. Next year the lawn will have patches of purple, red, yellow and white among the green. And the bees will have a feast. It will be beautiful.

Past the longest day and it will be only a few weeks until everything starts to die down. Flowers are already going to seed, roses to hips, leaves to earth. The season is turning but it's only just begun.

✿❀✿

Xena, Diplo and Chaos perch on the roof of Sussex Heights. Their parents are around, all five of them call to each other while shoppers and pub-goers mill beneath them on busy roads. They've well and truly fledged now. I think they're practising flying; one by one they launch from the roof, do a little turn and fly back. Baby steps.

They're so big and powerful, so majestic and so cute. I stand at the top of Regency Square with my binoculars, ignoring the bemused stares of passers-by. You know they've got their own website, says a man smoking on his doorstep. I turn to look at him and say yes. I want to tell him these three birds starting their lives on a tower block in central Brighton have been the only thing that's kept me going in the last three months but I think better of it. I say yes, yes, and smile, and get back to my birds. They continue to fly above me in little baby circles, their large, angular wings effortlessly launching them skywards. Their wing beats scarcely seem necessary, camp, almost. I wonder how these babies will go from this to being the fastest animal on the planet, when they will do their first dive. Will it be for a pigeon? I see them in the suburbs, sometimes. If I sit in the flat with the back door open, the gulls will tell me when the peregrines are here. Every one of them rises up from its rooftop nest and flies in agitated circles, cawing loudly. Pigeons scarper, house sparrows fall silent. Whole streets react in this way, gulls screaming, pigeons running, sparrows hiding, until the intruders leave and everything returns, instantly, to how it was. I've never been lucky enough to see a dive.

One of them flies to a rooftop car park and calls from there; pigeons and house sparrows scatter beneath them but they know it's not feeding time. They're cautious but they know they're safe. A hierarchy, a food chain played out on the mean streets of Brighton. Five birds own the city.

She was coning by the time her GP was called, he tells the student doctor sitting in on her routine appointment as she tries to convey her latest list of haemorrhage- or hospital-related ailments. I look it up: it means herniation of the brain. It means there was so much swelling inside the skull that her brain had started to be pushed where brain tissue isn't supposed to be. Coning. It's usually impossible to survive. I feel sick but I should feel lucky. I can't speak. I should feel grateful that she's not dead or brain-dead, that she doesn't have locked-in syndrome and that she can move all of her limbs. But I just feel sick. She had started to die.

It's still so impossibly early to know how well she'll recover. She's started cooking again, given up the bars of chocolate and whole packets of fish fingers she seemed so wedded to when she first got home. She's preparing veg, getting back to her routine. I call her and she tells me about the garden. Her speech is better but inconsistent. She wants me to come up and do some sausaging, she doesn't like sausaging alone. I'm confused for a minute before I realise sausaging is gardening. She wants me to weed the bloody veg patch and do something with the ornamental quince she tried to train up into an obelisk ten years ago and which still just sprawls out of the bottom. My fault for ignoring them while she was in hospital. My fault for leaving the worst jobs until last. Who knew she would come out of hospital having nearly died and be so concerned with sausaging? Mother of mine, I should have known.

I buy chips on the way home from watching the peregrines and let myself though the flat into the garden. I sit on the grass next to the pond forking salty, vinegary fingers into my mouth. I close my eyes, tune into bees. There are bumblebees still on the wing, the tinny whine from within a honeywort flower versus the low even buzz of flight. It's been a sunny day and insects abound. The back fence takes the last of the sunshine as the sky begins to pink. Herring gulls clack-clack into the fading light. The baby is at its rooftop sentry, its lone parent looks the other way. I can hear the house sparrows but I don't know where they are.

I finish my chips, ball the paper up and shove it in the compost. I fish out the secateurs and a little Tupperware box, cut down spent foxglove stems and shake the seeds into the container. Some end up scattering themselves on the ground and others are helped over the fence. But there's a fair amount of seed for raising more plants if the others fail, for working into other gardens, for creating more habitats.

We've come full circle, the garden, my mum and me. All of us broken, none of us quite fixed. But we'll get there, as Mum always says, we'll get there. Until the next thing happens, the next bucket of cement is poured over the living.

Once upon a time I lived in London. I rented a basement flat for six months while Jules and I looked to buy. The 'garden' was a patio – front and back. It was bare when we arrived but I grew tomatoes and sunflowers in pots, composted in a bucket. In the back yard we found seven frogs living in the drain. I asked the neighbour upstairs if he knew why they were there and he told me he'd filled in his garden pond when his daughter was born. Those frogs in the drain must have been searching for moisture in the hot summer, or perhaps they had grown from spawn laid there in desperation – they were all babies. The drain took water from the neighbour's kitchen above, which would often be coloured from paint from his daughter's play time, or contain soap from the washing machine or dishwasher. I made a pond for the frogs using a tub trug and a flower pot, which I placed inside, upturned and weighed down with a stone. Packet-bought watercress made adequate plant cover. The frogs remained fond of the drain, but one by one they found the tub trug. They would sit on the upturned pot and catch flies, and then at dusk disappear down their bespoke flight of steps to their evening quarters. They seemed brave for frogs, unlike others which shy away from humans, plopping into the pond at the first sight or sound of us. But then it was pointed out to me that, in a bare strip of patio, with no plants or anything else for shelter, there were few places to take refuge – they merely appeared brazen but they probably weren't. I made all sorts of piles for them then, using imported leaves and sticks, so they could hide. We saw less of them after that. But oh, did we love them. One of the frogs was bigger than the others and had a prominent hump on his back. We called him Humpy Back. In summer evenings I would take a bath with the back door open and listen to the slap-slap of Humpy coming in and exploring the bathroom and hallway.

The three-year-old daughter, for whom the pond was filled and the paint washed down the sink, ironically was obsessed with nature. She would beg her parents for a leaf or flower from my 'garden' whenever she went by, and I was happy to oblige. She liked tomato leaves best, soft and fragrant, easy to crush in a small hand. When we moved I gave her a sunflower that had managed to grow in a crack in the wall, and which the letting agents had forced me to remove to get our deposit back. The stem was prickly to touch and it towered above her but I like to think it opened a door to an unknown world she might one day want to explore.

When we moved it was only around the corner, and so I took Humpy Back and his friends with me. I couldn't trust any new tenants to keep a container pond going. I gathered them in a Tupperware box and transported them in a borrowed supermarket shopping trolley, along with the tomato plants, sunflowers and mobile compost bin – we didn't have a car and couldn't afford to hire a van. All survived the move and settled in well. I upgraded their pond to an antique tin baby bath, to which I added native plants to complement the watercress. The new garden backed onto the living room, so I would lie on the settee with the back door open, listening to the slap-slap of Humpy on the wooden floor. Sometimes, when it rained, I'd make myself a nest in front of the window and watch four or five of them on the patio, shoving worms into their mouth.

I think about Humpy Back a lot, about how long that pond had lain untouched in the garden above the basement I found him in, how many generations of frogs before him were able to breed successfully yet suddenly weren't able to. How unlikely it is that another garden would have a pond in that built-up area of London, that someone would come along and dig one. The garden has already been churned up, already a fence separates it from the basement. And if the house were further divided into flats, would a fence be erected, apportioning different 'outside spaces' for each floor? Fencing, decking, fake turf, bean bags, a 'garden room'. No habitat for frogs or means for them to reach it. Nothing green, nothing useful. How many years does it take to destroy a garden? For a population of frogs to die out? How many generations of chipping away so that, by the time everything has died, no one really notices?

In the 1980s my dad would clean insects off the bonnet, windscreen and headlights of his car. He did this religiously, worried the bloody, splattered carcasses would ruin his precious paintwork. When did you last clean insects off the bonnet of your car, Dad? Thirty years ago, he says.

Thirty years ago, our insect populations had already suffered a crash, after the drought of 1976 caused caterpillar food plants to shrivel and die and stopped flowers producing nectar. The ecosystems were already depleted, already broken, and some insects have never built their numbers back up since. We naturalists hark back to our childhoods as some rose-tinted vision of an insect-rich past. We are all David Attenborough talking about his buddleia. But in reality we are harking back to damaged goods, to things not right, to sustained, aggressive declines. We write about averages but we barely know what they are.

Thirty years before that my mum and her siblings played in a garden, around the corner from where I grew up. They found woolly bear caterpillars and collected frogs in buckets. They broke into the field at the bottom to ride Tiny the pony, and beyond that into the mere to collect frogspawn, sticklebacks and leeches. By the time I came along the garden had become a car park. Thirty years ago my mum would take me on walks around her Memory Lane, me on my bike, my Spokey Dokeys clattering noisily as I practised wheelies and tried to run over my sister, blissfully unaware of Mum's loss.

Thirty years before that, clouds of pipistrelle and Daubenton's bats would rise at dusk from the bridges of the Thames, black clouds of them hurtling around insects in the fading light. Now? 'Under favourable conditions,' writes Sladen in *The Humble-bee* in 1912, 'humble-bees store honey, the flavour of which, as most schoolboys know, is excellent.' Do they?

The further back we go the more wildlife we find. In Aix-en-Provence in southern France in 1608, residents found 'blood' on the walls of buildings and the local cemetery. Many believed it to be the work of the Devil but some suggested it was caused by butterflies. Further back in time, in 1553, naturalist Philip Henry Gosse described in his book, *The Romance of Natural History*, that hedges, trees, stones and people's clothes were sprinkled with 'drops of red fluid, which was supposed to be blood, til some observant person noticed the coincident appearance of unusual swarms of butterflies.' Some summers I gather caterpillars from stinging nettles and raise them in my kitchen. When they emerge from the chrysalis, they appear to bleed; it's a tiny amount, nothing more than a droplet. This bright red, blood-like liquid is meconium, described as 'a metabolic waste product from the pupal stage that is expelled through the anal opening of the adult butterfly'. Nice. How many butterflies would be needed for a whole village to think the Devil was at work? More than I will ever see in my lifetime. More than my great-grandparents ever saw in their lifetimes. It was called Blood Rain. It means something completely different now.

Thirty years ago I grew up in a garden that shaped me, made me. I played with moth cocoons, worms, pigeon feathers. I gathered runner beans and sneaked up on baby birds while their parents gathered food for them. I was photographed in front of naff 1970s shrubs and colourful butterflies. I sat among ants and made collections of leaves. I threw mud pies at my sister, climbed trees, learned to love.

In thirty years' time many children of today will not remember insects because they did not know insects. They will not remember butterflies because they did not know butterflies. They will look back on indoor childhoods, on screens and social media, on merchandising, on the park being too unsafe or rundown or unknown to play in. Insects will become extinct without them ever being seen, a door closing another inch on the natural world.

Every thirty years a little bit of land is ramped down, paved over. A new driveway, a garden office, a five-bedroomed house built in the gap between two others. A little more land is locked up, a new fence erected. Fences that deny children the right to explore and make mischief, fences that stop hedgehogs in their tracks. Like a waterfall, through which everything can travel freely from A to B, our gardens are gradually freezing over as winter sets in. We, the people, are winter. Will there ever be a thaw?

Each generation of naturalists harks back to their childhood, baselines shifting every thirty years. There was more wildlife in the Good Ol' Days, but there was also more squalor, cholera, infant mortality, feudalism and a deep-seated fear of the monarchy and God. Can we have enlightenment and health and insects? Can we have Democracy, Queer Rights and Women's Rights and insects? Must witches be burned at the stake for us to have insects? Can we reverse what we've broken without reversing what we've fixed?

Thirty years ago I read *Tom's Midnight Garden* and now I'm reading it again. I cry for the lost garden, the friendship between old and young, the shared love of a beautiful garden surrounded by fields that was long ago carved up to make houses and roads and driveways and creosoted fence panels. The garden will always be there, says Hatty, says Tom. Only one of them knows it would last only in dreams.

Thirty years ago I had the opportunity to ask my granny questions but I didn't. Instead she taught me things I took for granted then, but which I hark back to now. Since the morning she terrified me, in 1989, I've not heard a dawn chorus to match the one I heard with her. I wonder if it's as loud there now. It won't be.

I don't know what my garden looked like thirty years ago. It may already have been decked, depleted. Thirty years before that it might have been beautiful. A lone box bush, once probably clipped to a ball, suggests others might have existed in the space. The house was built in 1875. Before that it was fields. Before that woodland. At one stage there was a dairy and a baker's and, until quite recently, a farm. I wonder what lived and roamed here, and what will do so thirty years from now.

Jacques Cousteau said people protect what they love. But in thirty-year cycles we are forgetting what we love. This genetic imprint of woodland, wildflowers and the animal kingdom that we rely on for health and happiness and memories, which makes us human, is gradually being eroded. And with it, so is our wildlife.

A bumblebee lands on the wall, stops awhile, combs the hairs of her thorax with her little brush legs. She's a red-tail, all crushed-velvet black coat and bright-orange bottom. She's perfect and beautiful. I pretend she's a descendant of long-gone Adrienne, coming to tell me they're OK and thank you after all. I take her as a sign, a talisman. It's a warm day and I wonder if she's hot, busy from gathering pollen for her siblings. I check the bird bath for water, make sure there's a stone in it for bees and wasps to rest while they drink. We sit together in the sun, she combing her hair, me watching.

Adrienne's daughter stops combing, settles, as if thinking, as if pondering what to do next. I suppose I might have a drink, she says, I suppose I should get back to work. She half-heartedly gets up and buzzes, lazily, to the nearest borage flower, has a drink of nectar, moves on. She buzzes among other bees, from the borage to the honeywort, the chives to cranesbills. Only the last of these blooms remain, globe artichoke and agapanthus almost ready to take over.

She returns to her little bit of wall, bit of shade, returns to combing. I wonder where her nest is. An abandoned garden maybe, a bit of brownfield land, an allotment. Perhaps celebrated in a well-tended and loved garden – we've got bees! Yes! I hope so. Red-tails can nest in walls, under sheds, in old mouse holes, an old duvet thrown out into the yard. If you build it they will come, if you don't they may come anyway.

But it would matter hugely if we gave them a helping hand. If we stop razing gardens and paving them over, drowning what's left with poison and suspicion. What could this land become if we just let things be? If we learned to love a bit more, to let things work themselves out. If we outlawed slate chippings and weed-suppressant membrane? If we grew out of this horrid obsession with 'outdoor rooms', paved-over front gardens, fenced-off land, convenience. How happier we would be – look after the bees and everything will follow. Everything including us.

Mum called me today. For a chat, she said. She's not eating enough vegetables and she wants me to bring some from the garden on my next visit. She wants to eat some fish. She's doing so well but I wish she could see it. She's 80 per cent fixed but sees 20 per cent still broken. She's beating her friends at Scrabble again, completing the hard Sudoku. She's amazing. The doctors and nurses are amazing. The resilience of living things is amazing. Today I understand half of what she says. I correctly guess a quarter and pretend to know the rest. I'll get there, she says.

The bee stirs again, engages her flight muscles, lifts herself up and over the flowers, higher and away as a helicopter rising into the clouds. As she ascends I imagine her view; the garden becomes smaller and more far away, the walls irrelevant, unseen, the divided patches of land as one. Off she goes into the blue, high above the houses with people crammed in them, the roofs and chimneys with the seagulls, the mossy gutters, the laughing starlings, the jackdaws. What does a fragmented, fenced-off land look like from up there? What would it look like if we all made our gardens better, joined them together? Gardens linked all the way to the train tracks, the train linking Hove to Brighton, to London, to Mum, to the garden, to every garden I have loved and lost and every bit of land that's ever been wild. Ever. Higher and higher she flies until all she can see is green, habitat-rich networks allowing all to travel, feed and breed. Life pulsing through Britain's veins.

Far below her now the woman sits on the lawn in her little parcel of recovered land. Grass sways in the breeze, flowers nod to lure bees. There are holly blue and speckled wood butterflies, a lone red admiral soaking up the sun. Leaves hide hoppers and miners, aphids and flies. Above the pond a second generation of common darter dragonflies dances for a mate. Life. It just needs a chance. We just need to give it a chance.

Species list

Below are all the species I saw and was able to identify in my garden. There are a few leafminers and other critters I couldn't work out, and the lack of moths and their caterpillars is more a reflection of where I was in spring and summer (Birmingham) than the actual number that visited my garden. But still, for a garden brought back from the dead, I'm quite pleased.

Birds
Blackbird *Turdus merula*
Blue tit *Cyanistes caeruleus*
Collared dove *Streptopelia decaocto*
Dunnock *Prunella modularis*
Feral pigeon *Columba livia domestica*
Goldfinch *Carduelis carduelis*
Great tit *Parus major*
House sparrow *Passer domesticus*
Robin *Erithacus rubecula*
Starling *Sturnus vulgaris*
Woodpigeon *Columba palumbus*

Bees and wasps
Blue mason bee *Osmia caerulescens*
Buff-tailed bumblebee *Bombus terrestris*
Common wasp *Vespula vulgaris*
Early bumblebee *Bombus pratorum*
Ectemnius wasps *Ectemnius* species
Furrow bee *Lasioglossum calceatum/ albipes*

Garden bumblebee *Bombus hortorum*
Gooden's nomad bee *Nomada goodeniana*
Hairy-footed flower bee *Anthophora plumipes*
Heath bumblebee *Bombus jonellus*
Honey bee *Apis mellifera*
Ichneumon wasp
Leafcutter bee *Megachile centuncularis*
Leafcutter bee *Megachile ligniseca*
Orange-vented Mason Bee *Osmia leaiana*
Plasterer bee *Colletes* species
Red mason bee *Osmia bicornis*
Red-tailed bumblebee *Bombus lapidarius*
Toothed flower bee *Anthophora furcata*
Tree bumblebee *Bombus hypnorum*
White-spotted sapyga *Sapyga quinquepunctata*
White-tailed bumblebee *Bombus lucorum*
Wool carder bee *Anthidium manicatum*
Yellow-faced bee *Hylaeus confusus*

Butterflies and moths
(NB: comparatively few records due to lack of a moth trap)
Angle shades *Phlogophora meticulosa*
Brown-tail moth *Euproctis chrysorrhoea*
Buff ermine moth *Spilarctia luteum*
Holly blue butterfly *Celastrina argiolus*
Large white butterfly *Pieris brassicae*
Peacock butterfly *Aglais io*
Red admiral butterfly *Vanessa atalanta*
Silver Y moth *Autographa gamma*
Small tortoiseshell butterfly *Aglais urticae*
Small white butterfly *Pieris rapae*
Wax moth *Aphomia sociella*

Flies
Bluebottle *Calliphora vomitoria*
Broad centurian fly *Chloromyia Formosa*
Common drone fly *Eristalis tenax*
Crane fly *Tipula paludosa*
Greenbottle *Lucilia sericata*
Hoverfly *Eristalis intricarius*
Hornet hoverfly *Volucella zonaria*
Lacewing *Chrysoperla carnea*
Long hoverfly *Sphaerophoria scripta*
Marmalade hoverfly *Episyrphus balteatus*
Hoverfly *Myathropa florea*
Narcissus bulb fly *Merodon equestris*
Non biting midge *Chironomidae* species
Hoverfly *Platycheirus peltatus*
Hoverfly *Syrphus ribesii*
The footballer *Helophilus pendulus*
Hoverfly *Volucella bombylans*
Hoverfly *Volucella inanis*
White fly *Trialeurodes vaporariorum*
White-footed hoverfly *Platycheirus albimanus*

Pond life
Common blue damselfly *Enallagma cyathigerum*
Common darter dragonfly *Sympetrum striolatum*
Diving beetle *Agabus bipustulatus*
Large red damselfly *Pyrrhosoma nymphula*
Pond snail *Lymnaea stagnalis*
Rose leaf miner *Stigmella anomalella*

Spiders and harvestmen
Garden spider *Araneus diadematus*
Giant house spider *Eratigena atrica*

Harvestmen *Opiliones*
Woodlouse spider *Dysdera crocata*

Mammals
Fox *Vulpes vulpes*
Grey squirrel *Sciurus carolinensis*

Other critters
Black ant *Lasius niger*
Black bean aphid *Aphis fabae*
Common centipede *Lithobius forficatus*
Common pill woodlouse *Armadillidium vulgare*
Common rough woodlouse *Porcellio scaber*
Common striped woodlouse *Philoscia muscorum*
Common woodlouse *Oniscus asellus*
Earthworms *Lumbricina*
Froghopper *Issus coleoptratus*
Leaf miner *Agromyzidae*
Mosquito *Culicidae*
Non-biting midge *Chironomidae*
Rose aphid *Macrosiphum rosae*
Rose sawfly *Arge pagana/Arge ochropus*
Rove beetles *Staphylinidae*
Willowherb aphid *Aphis epilobii*

Author acknowledgements

Thank you…

To my agent, Jane Turnbull and my editor, Julie Bailey, both of whom took a chance on my half-baked idea of writing a book about a garden I hadn't started. And to my copy-editor, Mari Roberts, who showed me how to write a book, and for her kindness and patience when I wanted to make so many last-minute changes after all the work she had done. To illustrator extraordinaire Jessie Ford, for my wonderful cover, and to Rachel Nicholson, Kealey Rigden, Hannah Paget and everyone at Bloomsbury.

To Helen Ginn, who not only gave me her greenhouse and lots of plants, a bumblebee nest and hours of cups of tea and chats but who also read my manuscript and offered unbiased, sensible advice, all while finishing her own book and looking after baby twins. You're amazing, really.

To Adrian Thomas for showing me where the nightingales are. To Michael Blencowe for sending me seeds. To Ben Darvill for always being on hand with bumblebee queries, to Twitter for being endlessly clever and entertaining.

To Andrew and Louise Donald, thank you for looking after the garden and for not putting a tennis court over it after all.

To Jo Wright and Pete Connolly for providing information about the garden at 86 when Mum wasn't able to.

To the doctors and nurses at Heartlands, Queen Elizabeth and Solihull hospitals for caring for Mum and saving her life. To the cleaners and porters who made her recovery more comfortable, to the therapists who helped her adjust back to the real world.

To my family for just about holding things together, and for allowing me to tell this story. To Trudi. To the red-tailed bumblebee that came into my life all those years ago, and to Jonny for throwing that smelly old duvet into the yard.

If you want to learn more...

If you want to learn more about, or create habitats for, wildlife in your garden, these are some of my favourite books, which you might like too. Some are out of print so you'll need to hunt them down in charity shops or online, but they're worth it. I've also included my first book, *The Wildlife Gardener*, a practical guide to creating habitats for wildlife in your garden, and overleaf I've listed websites for just some of the many organisations working hard to help wildlife.

Wildlife Gardener by Kate Bradbury (Bloomsbury, 2019)

Our Garden Birds by Matt Sewell (Ebury, 2012)

Organic Gardening by Charles Dowding (Green Books, 2013)

Gardening for Birdwatchers by Mike Toms and Ian and Barley Wilson (British Trust for Ornithology, 2008)

Hedgehogs by Pat Morris (Whittet Books, 2014)

Frogs and Toads by Trevor Beebee (Whittet Books, 1997)

The Secret Life of Garden Birds by Dominic Couzens (Bloomsbury, 2004)

The Secret Life of Flies by Erica McAlister (The Natural History Museum, 2017)

Field Guide to the Bumblebees of Great Britain and Ireland by Mike Edwards and Martin Jenner (Ocelli, 2009)

Field Guide to the Bees of Great Britain and Ireland by Steven Falk and Richard Lewington (Bloomsbury, 2015)

Butterflies of Britain and Ireland by Jeremy Thomas and Richard Lewington (Bloomsbury, 2014)

A Natural History of Ladybird Beetles by M.E.N. Majerus, H.E. Roy and P.M.J. Brown. (Cambridge University Press, 2016)

Beetles (Collins New Naturalist Library) by Richard Jones (Collins, 2018)

Field Guide to the Dragonflies and Damselflies of Great Britain and Ireland by Steve Brooks and Steve Cham (Bloomsbury, 2014)

Amphibian and Reptile Groups of the UK
arguk.org

Bat Conservation Trust
bats.org.uk

British Dragonfly Society
british-dragonflies.org.uk

British Hedgehog Preservation Society
britishhedgehogs.org.uk

British Trust for Ornithology
bto.org

Buglife
buglife.org.uk

Bumblebee Conservation Trust
bumblebeeconservation.org

Butterfly Conservation
butterfly-conservation.org

Flora locale
floralocale.org

Froglife
froglife.org

Hawk and Owl Trust
hawkandowl.org

Moths Count
mothscount.org

Red Squirrel Survival Trust
rsst.org.uk

The RSPB
rspb.org.uk

Swift Conservation
swift-conservation.org

The Mammal Society
mammal.org.uk

The RHS
rhs.org.uk

The Wildlife Trusts
wildlifetrusts.org

UK Ladybird Survey
ladybird-survey.org

Wild About Gardens
wildaboutgardens.org.uk

WWF UK
wwf.org.uk

Index